# Consumption and Public Life


Series Editors
Frank Trentmann
Birkbeck, University of London
London, UK

Richard Wilk
Indiana University
Bloomington, IN, USA

The series will be a channel and focus for some of the most interesting recent work on consumption, establishing innovative approaches and a new research agenda. New approaches and public debates around consumption in modern societies will be pursued within media, politics, ethics, sociology, economics, management and cultural studies.

Arne Dulsrud • Francesca Forno
Editors

# Digital Food Provisioning in Times of Multiple Crises

## How Social and Technological Innovations Shape Everyday Consumption Practices

palgrave
macmillan

*Editors*
Arne Dulsrud
Consumption Research Norway
Oslo Metropolitan University
Oslo, Norway

Francesca Forno
Sociology and Social Research
University of Trento
Trento, Italy

ISSN 2947-8227 ISSN 2947-8235 (electronic)
Consumption and Public Life
ISBN 978-3-031-46322-8 ISBN 978-3-031-46323-5 (eBook)
https://doi.org/10.1007/978-3-031-46323-5



This Palgrave Macmillan imprint is published by the registered company Springer Nature Switzerland AG.
The registered company address is: Gewerbestrasse 11, 6330 Cham, Switzerland

Paper in this product is recyclable.

# Foreword

Digital platforms—most visible in the applications downloaded onto smart devices and bookmarked in website browsers—have been met with diverse claims, both utopian and dystopian, regarding societal transformation. Claims about their role in facilitating food sustainability transitions are a case in point. We often hear how platforms promise supply chain reforms that facilitate more circular economies and alternative modes of provisioning through sharing economies (Llamas & Belk, 2013; Gensch et al., 2017) that will shake up the global food system. At the same time, fears abound regarding how the data underpinning such platforms will lead to targeted marketing, increase levels of consumption while also reinforcing inequalities and serving to further entrench the power of global food corporations (Singh & Lyon, 2013; Srnicek, 2017; Hazas & Nathan, 2018; Zuboff, 2019). Whatever the prognosis, prolonged periods of lockdown across many nations during the Covid pandemic have accelerated the transition towards digital platform-mediated consumption. If four years ago the rise of the platform was held as something to be carefully assessed and, perhaps, an opportunity to systematically foster sustainability transitions (Welch & Southerton, 2019), then today (2023) debates have already turned to concerns about how to understand the already embedded platformitisation (Talamini et al., 2022) of food provisioning and consumption.

This anthology is unique in addressing the key questions raised by the rise of digital platforms in the context of sustainable food. Already engaged in the delivery of a major project comparing conditions and experiences of digital food platforms across five European countries when Covid struck, the contributions to this book represent crucial early insights into online food platforms and their implications for sustainable food systems. The chapters are diverse. They offer conceptual delineation, underpinned by robust empirical analysis, of platform types. Legal implications and digital consumer rights are addressed, alongside investigations of food waste, consumer routines, questions of taste, and understandings of "good" food. Debates regarding the capacity of digital platforms in supporting alternative food networks receive consideration and demonstrate the complex relationships between ideas of environment, nature, sustainability, health, and community. Social relations, particularly in relation to gender, age, and socio-economic status, feature prominently in the analysis. Attention is given to the theoretical frameworks that are mobilised in seeking to address critical questions of societal change, and the methodological appendix provides an essential resource for the further development of empirical investigations. This anthology does not provide the final say on digital platforms or sustainable food transitions; it provides the critical social science resources for research to answer the pressing questions that these topics raise.

University of Bristol Business School
University of Bristol, Bristol, UK

Dale Southerton

## References

Gensch, C.-O., Prakash, S., & Hilbert, I. (2017). Is digitalisation a driver for sustainability. In T. Osburg & C. Lohrmann (Eds.), *Sustainability in a digital world: New opportunities through new technologies* (pp. 117–129). Springer.

Hazas, M., & Nathan, L. (2018). *Digital technology and sustainability: Engaging the paradox*. Routledge.

Llamas, R., & Belk, R. (2013). Living in a digital world'. In R. Belk & R. Lamas (Eds.), *The Routledge companion to digital consumption* (pp. 3–12). Routledge.

Singh, S., & Lyon, D. (2013). Surveilling consumers: The social consequences of data processing on amazon.com. In R. Belk & R. Lamas (Eds.), *The Routledge companion to digital consumption* (pp. 319–332). Routledge.

Srnicek, N. (2017). *Platform capitalism*. Polity Press.

Zuboff, S. (2019). *The age of surveillance capitalism*. Profile Books.

Talamini, G., Li, W., & Li, X. (2022). From brick-and-mortar to location-less restaurant: The spatial fixing of on-demand food delivery platformization. *Cities, 128*.

# Acknowledgements

This book is a result of evolving series of discussions taking place among the researchers of the project "Sustainable Food Platforms: Enabling Sustainable Food Practices Through Socio-technical Innovation (PLATEFORMS)". Evolving, because events that occurred during our project (COVID-19) affected both the direction and the framing of our inquiry. Although the preventive actions against COVID-19 came in the way for our methods and progress, the pandemic also demonstrated the relevance of our endeavour. The access to food online became an aliment and a necessity for a vast number of households under isolation and thereby becoming an integral part of consumption practices.

We would like to acknowledge and extend our gratitude to several individuals and institutions towards making this volume. This research was made possible by funding from SUSFOOD2, the ERA-Net Cofund on sustainable food production and consumption co-funded under European Union's Horizon 2020 research, and innovation programme under Grant Agreement n°727473. The SUSFOOD 2 secretariat created a fertile ground for discussions through annual seminars and workshops which provided us with valuable input and encouragement. Funding institutions at the national level have played an important role in facilitating and supporting the national research teams. These were the Research Council of Norway (RCN), Swedish Research Council for Sustainable Development (FORMAS), the Department of Agriculture, Food and the Marine (DAFM) in Ireland, Ministry of University and Research (MUR) in Italy, and Federal Ministry of Food and Agriculture (BMEL) in Germany. There are colleagues who made significant contributions to the content of this book, but who are not represented as authors. A particular gratitude goes to colleagues Marie Hebrok and Nina Heidenstrøm at Consumption Research Norway and Emanuela Bozzini at the University of Trento. Professor Dale Southerton at our advisory board followed our project closely from the very beginning; he commented drafts, edited texts, and opened the doors to other publication outlets.

Last, but not least, we owe debt to Sharla Plant at the Sociology and Social Policy section at Palgrave Macmillan who took the initiative to publish our research. Her friendliness encouraged us, and her patience during the pandemic gave us the time to develop and improve our texts.

# Contents

# Notes on Contributors

**Frode Alfnes** is Professor of Economics in the School of Economics and Business at Norwegian University of Life Sciences, Norway. He is also an Adjunct Research Professor at Consumption Research Norway, Oslo Metropolitan University (OsloMet). Alfnes' research focuses on a variety of topics related to consumer behaviour and economics, including sustainable consumption, food markets, green energy transition, and behavioural and experimental economics.

**Silvia Biasetton** is an early career researcher and MA graduate in Sociology and Social Research from the University of Trento, Italy, and Cultural Sociology from Masaryk University, Czech. Her research interests range from cultural sociology, sociology of food and consumption, to cultural anthropology, sociology of sexuality, gender, and post-human studies.

**Suse Brettin** is a PhD student in Agricultural Sciences. She is currently working on a feminist perspective on practices of care in the context of agricultural production. Furthermore, her research focuses on human-nature relations in the fields of agriculture, food, and water, as well as on the analysis of alternative value chains and food networks.

**Meike Brückner** is a postdoctoral researcher in the Division of Gender and Globalisation at Humboldt-Universität zu Berlin, Germany, where she researches and teaches on the practice and politics of food and agriculture, human-nature-relations, digitalisation and sustainability, and qualitative and ethnographic methodologies. She has completed her PhD on Biodiversity, Local Crops, and Food Sovereignty in Kenya.

**Margherita Brunori** is Assistant Professor of Food and Agriculture Law in the Department of Law at the University of Brescia, Italy. She has an interdisciplinary research background, thanks to her postdoc experience in the Department of Sociology and Social Research at the University of Trento and in the Department of International, Legal, Historical, and Political Studies at the University of Milan. She is a specialist in the international, European, and national governance of food systems, rural development, and access to natural resources.

**Sandra Cajic** is a researcher and MA graduate in Gender Studies from the Humboldt-Universität zu Berlin, Germany.

**Jiaxi Chen** is a doctoral student in the University of Bristol Business School, UK. His research interests lay in the intersection of digital platforms and everyday food consumption. He is also interested in how digital technologies are changing people's everyday life, with a theoretical focus on social practice and the use of qualitative research methods.

**Alice Dal Gobbo** is an assistant professor in the Department of Sociology and Social Research at the University of Trento, Italy. Her research interests lay at the intersection of political ecology and the sociology of everyday life, particularly in the fields of food and energy. She is also interested in reflections and experimentations on qualitative methodologies.

**Arne Dulsrud** is a Research Professor at Consumption Research Norway (SIFO), Norway. He has written widely on consumer policy issues, economic sociology, and food policy both in books and in journals including *Journal of Consumer Policy*, *Consumption and Society*, and *Acta Sociologica*. From 2018 to 2021, he coordinated the Eranet-SUSFOOD2 PLATEFORMS PROJECT, aiming to produce in-depth and practically oriented knowledge on how sustainable food practices can be enabled through socio-technical innovations in food provisioning platforms. He is now coordinating "Digital Infrastructures for Sustainable Consumption: Redirecting, Reorganizing, Reducing and Reimaging Consumption (DISCo)", a three-year project including five countries granted by Belmont Forum Systems of Sustainable Consumption and Production (SSCP 2022).

**Francesca Forno** is Associate Professor of Sociology at the University of Trento, Italy. Her research interests include political consumerism and sustainable community movement organisations. She has written articles and books on civic participation and social movements, conducting research on political consumerism, collaborative consumption, grassroots initiatives on social eco-innovation, and alternative food networks. Her work has appeared in leading international journals including the *Annals of the American Academy of Political and Social Science*, *The Journal of Consumer Culture*, *The International Journal of Consumer Studies*, *The British Food Journal*, and *Food Policy*, as well as in collections of essays.

**Christian Fuentes** is a professor in the Department of Service Studies. His research focuses on topics such as digital food platforms, ethical consumption, alternative markets, sustainable retailing, mobile shopping, and e-tailing. He is currently involved in various projects that explore the role of digital platforms in enabling sustainable consumption.

**Maria Fuentes** is an associate professor in the Department of Service Studies at Lund University, Sweden. Her research focuses on the way everyday food consumption is enabled and shaped by marketing practices and market artefacts. She has taught and conducted research on the topics of food consumption, food risk, alternative markets, and novel food.

**Shadi Hashem** was a postdoc researcher on the PLATEFORMS project (2018–2021) in the Cork University Business School at University College Cork, UK.

**Jenny Ivre** holds a Master's Degree in Service Management, specializing in retail, from the Department of Service Studies, Lund University, Sweden. She wrote her master's thesis on the Karma food app in connection to the PLATEFORMS project. She currently works as a talent acquisition consultant at Netcompany.

**Natalia Magnani** is Associate Professor of Environmental Sociology and Sociology of Territory in the Department of Sociology and Social Research at the University of Trento, Italy. Her research interests include the socio-ecological transition in the energy and food systems, sustainable development, urban and rural development, and environmental conflicts. On these issues she has written articles in various international journals such as the *Journal of Rural Studies*, *Sustainable Production and Consumption*, and *Energy Research & Social Science*, as well as two monographs and chapters in national and international volumes.

**Mary McCarthy** is Professor of Marketing in the Cork University Business School at University College Cork, UK. Mary's research interests lie in the field of consumer behaviour, with particular reference to food. She explores the social, cognitive, and economic factors that influence consumer decision processes and behaviours.

**Chris Moran** is an Irish Research Council-funded PhD scholar at University College Cork, UK, and a research assistant on the Irish team of the PLATEFORMS project. His PhD research focuses on food consumption and sustainability.

**Filippo Oncini** is a JSPS fellow in the Department of Japanese Food Culture at Kyoto Prefectural University, Japan, and Honorary Research Fellow at the Sustainable Consumption Institute, University of Manchester, UK. His research interests lie in the fields of consumption and cultural sociology, with attention to social inequalities, food consumption and provision, and urban poverty. His recent projects focus on the relations of cooperation and competition between different food charities in the UK and Japan using field theory. Since October 2024 it has joined the Maastricht Sustainability Institute at Maastricht University as an assistant professor.

**Claire O'Neill** is Lecturer in Marketing in the Cork University Business School at University College Cork, UK. Claire's research focus/projects are linked to SDG 12 and focus on sustainable consumption/sustainable consumer behaviour in the areas of waste, energy, water, recycling, and food.

**Emma Samsioe** is an assistant professor in the Department of Service Studies at Lund University, Sweden. Her research focuses on fashion and food consumption. Much of her work has explored the role of digitalisation in enabling more sustainable modes of consumption. She is currently working on various projects on the topic of circular fashion consumption.

# Abbreviations

| | |
|---|---|
| AFNs | Alternative Food Networks |
| ANT | Actor-Network Theory |
| CSA | Community-Supported Agriculture |
| EA | Early Adopters |
| EU | European Union |
| FA | Food Assembly |
| GAS | Gruppi di Acquisto Solidale (Solidarity Purchasing Groups) |
| IB | Impulse Buying |
| ICTs | Information and Communication Technologies |
| IT | Information Technology |
| MLP | Multi-Level Perspective |
| NCDs | Non-Communicable Diseases |
| NRU | Not Regular Users |
| OFPS | Online Food Provisioning Services |
| SPT | Social Practice Theory |
| STS | Science and Technology Studies |
| ToP | Theories of Practice |

# List of Figures

**Online Food Provisioning Services and Where to Find Them: Pipelines, Platforms and the Rise of Dark Stores**



**From Grassroots to Platforms: How Digitalisation Reconfigures Learning and Engagement with Food**

**Online Food Provisioning Services (OFPS) as Sociotechnical Assemblages: Three Theoretical Threads and a Proposition for Further Research**

# List of Boxes

# List of Tables

# Digital Food Provisioning in a Time of Multiple Crises: An Introduction

*Arne Dulsrud and Francesca Forno*

## 1 Introduction

This anthology examines the rise of digital platforms for food provisioning and explores their impact on the sustainability of consumption practices. The central hypothesis of the book is that how food is purchased influences how it is consumed. Thus, not only the location where food is eaten (e.g. households, workplaces, restaurants) but also the method of its acquisition affects how food is prepared, consumed, and discarded.

The complex process through which food reaches the consumer's table has become a central topic in political and public debate. This discussion

A. Dulsrud (✉)
Consumption Research Norway, Oslo Metropolitan University, Oslo, Norway
e-mail: arned@oslomet.no

F. Forno
Department of Sociology and Social Research, University of Trento, Trento, Italy
e-mail: francesca.forno@unitn.it



A. Dulsrud, F. Forno (eds.), *Digital Food Provisioning in Times of Multiple Crises*, Consumption and Public Life,
https://doi.org/10.1007/978-3-031-46323-5_1

has also been pushed by the growing awareness among citizens of the negative social and environmental impacts of the conventional agrifood supply chain and their increasing interest in the healthiness of their diets. Indeed, the production, distribution, consumption, and disposal of food have significant environmental impacts, such as soil degradation, water and air pollution, climate change, and biodiversity loss. Therefore, developing sustainable forms of food production, distribution, and consumption is crucial to achieving an ecological transition (Alfredsson et al., 2018).

Over recent years, the international community has gradually acknowledged that the way food systems are evolving needs to be coupled with sustainability. One important example is the contribution of the current food system to climate change: about 21–37% of total greenhouse gas (GHG) emissions are attributable to the food system (IPCC Special Report Climate Change and Land, 2019). The societal and human impacts are similarly severe: two billion people suffer from obesity and related diseases, while 836 million people are still undernourished, although annually one-third of the food produced—about 1.3 billion tonnes per year—is wasted (Gustavsson et al., 2011). More recently, the so-called Triple Cs—COVID, Conflicts, and the Climate Crisis—have worsened an already critical situation. These events, among others, have put the need for a resilient and sustainable food system both on the EU's political agenda through the Green Deal and on the programme of the United Nations Climate Change Conference (COP27) where the need to accelerate innovation and digitalisation in the food sector to combat climate change is being heavily emphasised.

Digital technologies, many argue, contain a great promise and an opportunity for more sustainable consumption of food in everyday life (Cooper et al., 2012; Graham & Haarstad, 2011; Nemes et al., 2021). Digital platforms can bypass conventional and slow-moving retail supply chains through direct distribution. They provide opportunities for innovation niches and circular food distribution systems. The various applications can assist consumers to choose climate-friendly products in-store. In addition, digital platforms can provide sharing options for surplus food and stimulate networks between retailers and consumers. They can enable the formation of ethical consumption communities which exert a pull effect in the value chain (Schneider et al., 2018).

Based on this thinking, it is reasonable to ask whether digital technologies can provide the expected paradigmatic shift to a more sustainable consumption pattern. Southerton and Fuentes (2022) argue that to realise these potentials, digital food platforms must be able to 'reconfigure' (Geels et al., 2015) what are often assemblages of household food practices, held in place by clusters of socio-technical arrangements, norms, and habits (see also Samsioe & Fuentes, 2021). Major research questions are raised concerning what is required to make such reconfigurations possible and the sustainability implications of those changes. At the same time, as more critical scholars argue, digitalisation often also leads to increased surveillance, the misuse of personal data, unequal access, manipulation, and the acceleration of unfair material distribution (Zuboff, 2019; Srnicek, 2017; Dulsrud & Bygstad, 2022).

In recent years, digitisation has coincided with a growing number of so-called alternative food networks (AFNs). As demonstrated by the literature on AFNs (Goodman et al., 2012; Sonnino & Marsden, 2006; Forssell & Lankoski, 2015; Bos & Owen, 2016; Corsi & Novelli, 2018; Fuentes & Fuentes, 2022), these new forms of provisioning offered to consumers provide an alternative to the mass retail business model by counteracting the negative consequences of conventional food production, consumption, and distribution (Bos & Owen, 2016; De Bernardi et al., 2019). AFNs seek to reorient economic power structures in response to various food system-related issues not just concerning consumer health but also environmental impacts. The increase in digital platforms and online shopping provisioning services (e.g. box schemes, so-called food assemblies, etc.) has led to the multiplication of new supply network initiatives. It should be noted, however, that digital food platforms should not be seen as a direct evolution of physical AFNs, as only in some cases is there a link between grassroots physical food platforms and online food provisioning services, the latter often being commercial in nature (Oncini et al., 2020). Each platform connects producers and consumers in new and innovative configurations, presenting users with a unique choice architecture.

As the food supply landscape is rapidly transforming in a time of crisis and rapid socio-technical innovation, it is crucial that the social sciences can relate to these changes and provide thoughtful accounts of their social and environmental consequences. This book brings together a carefully curated range of insights from a team of researchers from different fields (sociology, marketing, food and gender studies, economics, and law) working in different European universities who were engaged in a collaborative project for more than three years. Grounded in empirical research conducted in Norway, Italy, Germany, Ireland, and Sweden, this anthology explores how (and if) socio-technical innovations in food provisioning are fostering new and more sustainable forms of food consumption in everyday life, investigating if and how digital food platforms can drive/facilitate sustainable provisioning practices in households.

## 2 Bringing the Consumer into Platform-Based Economy Research

The recent and expanding discourse on new online food provisioning services has generally portrayed the opportunities provided by new digital platforms as more sustainable, often in an uncritical manner (Michelini et al., 2018). Social media, apps, direct selling websites, and food platforms have often been considered unambiguously positive sustainability tools. Typically, three key aspects have been emphasised in both public and academic literature. Firstly, it has been claimed that digital platforms can help small- and medium-sized producers sell their products at a fairer price while enabling consumers to access quality food less expensively than in large retailers due to reduced supply chain costs. Thus a win-win situation results that promotes food sustainability in its three dimensions: environmental, social, and economic (Huyghe et al., 2017).

Furthermore, it has been argued that digital provisioning channels can positively contribute to sustainability by providing access to fresh and local foods that would otherwise be difficult to reach, especially in urban contexts, and/or by enabling the purchase of produce that would be less easy to find or learn about, such as niche products (Fuentes & Sörum, 2019; Dal Gobbo et al., 2022). For example, if one is interested in purchasing fair trade products, online food provisioning services could facilitate access to them while providing more information about their ethical and environmental characteristics (cf. Seyfang, 2005).

Finally, another often highlighted aspect of these online services concerns waste reduction. There are many examples of platforms that allow individuals, restaurants, or retailers to make surplus or near-expiry food available at a cheaper price (Davies & Legg, 2018; Fuentes et al., 2021). Digital platforms could also contribute to better planning of what is produced and what is purchased, thus reducing transport and other energy-intensive forms of food storage/preservation—in other words, reducing waste 'behind the shelf' on both sides of the food supply chain (De Bernardi et al., 2019).

However, despite the increasing availability and popularity of digital food platforms, a better understanding of their impact on consumption practices is still needed to determine whether they represent a genuine opportunity to promote more sustainable diets. Moreover, it is worth noting that most studies not only hypothesise that food platforms could provide a more sustainable option but also tend to underestimate the issue of 'recruitment'. Exploring how these new sales channels can effectively reach and retain consumers is a crucial aspect that requires further investigation, also in view of the fact that online food shopping remains less popular than other retail channels and this despite the boost received during the COVID-19 pandemic (Oncini et al., 2020; Holmberg et al., 2022; see also Oncini in this volume).

Logistical aspects certainly make investments in online food retail more complex and risky than in other sectors (books, clothing, etc.), which might explain why, for example, new food movement organisations have only partially taken advantage of the potential offered by digital platforms. Still, it is essential to assess in more detail to what extent these new digital tools can become part of the everyday eating practices of consumers and how they may enable, in reality, more sustainable consumption. In other words, it is necessary to deepen our understanding of the dynamics of the digitisation of food from a consumer perspective.

Indeed, while motivations for healthier and more sustainable eating may explain consumers' inclination to try new shopping channels, it is important to recognise that adopting a new shopping routine is not an easy task and may require time, resources, and/or the acquisition of new skills. Furthermore, some digital platforms may be more suited to certain individuals than others. For example, platforms that deliver food at home (such as 'digital box schemes') are more amenable to those with stable jobs and fixed working hours (Hertz & Halkier, 2017). Consequently, the ability of digital shopping channels to attract consumers may vary according to people's lifestyles and skills.

In other words, if food consumption is important for shaping the sustainability of the food supply chain, transitioning to a more sustainable food system requires understanding and active engagement with households and their multiple food practices. In addition to food acquisition, it is equally important to understand how food is handled once it enters the household. Put differently, to assess the potential of new digital food platforms to promote sustainable eating practices, it is necessary to comprehend how everyday food practices fit within household life and how the interaction with new supply services becomes integrated into these complex arrangements, supporting or hindering consumers' sustainable practices (Fuentes & Samsioe, 2020).

## 3 The PLATEFORMS Project

The PLATEFORMS project aimed to produce in-depth knowledge regarding how food practices are affected by how food enters the home, with an overall objective of producing deep, practical knowledge on how and if more sustainable household food consumption can be enabled through innovations in food provisioning platforms. We asked: can digital food platforms make consumption more sustainable, and if so, how? How can we advise platform owners and retailers about how to facilitate sustainable consumption practices?

Grounded in empirical research conducted in five European countries, the project embraced a comparative and interdisciplinary approach to explore—through an examination of multiple digital food platforms—if and how new online purchasing channels reconfigure household practices across different cultural and social contexts.

### *The PLATEFORMS methodology*

Our research questions were explored through various methods and at different analytical levels. At the macro level, we undertook a systematic mapping of online food platforms in all five countries. We had a broad approach, including mass retail platforms and food box schemes: from Amazon to local food cooperatives.

At first, desk research was undertaken to contextualise the food provisioning structures in Ireland, Italy, Sweden, Norway, and Germany. Following this, an audit of Online Food Provisioning Services (OFPS) was undertaken. Analysis of the OFPS websites followed a two-stage process. Stage one involved the development of a standard codebook for categorising information extracted from the OFPS websites. After the initial screening and deletion of duplicates, 524 websites were identified as OFPS across all five countries. The second stage of analysis involved coding and verifying these data. The results of this audit considered the structure and value propositions communicated by these provisioning services. The catalogue of online platforms was coded and analysed according to several criteria addressing healthy food and sustainability indicators. The results provided an overview of the structure and value propositions communicated through these provisioning services (see Appendix A).

Secondly, we conducted a household study to explore the nexuses of social practices that surround and frame sustainable consumption. The project employed an ethnographically inspired approach based on social practice theory (SPT). This 'practice-centred' framework focuses on the 'doings' of consumption to highlight the nature of household routines and everyday engagements with food. A sample of more than 40 households per country, selected within four cases (Alternative physical, Alternative digital, Mainstream physical, and Mainstream digital), was used to explore household food consumption practices from acquisition to disposal. A photographic 'food diary' was recorded by participants and was followed by an in-depth interview regarding their food consumption practices from acquisition to disposal. This was supplemented with a kitchen 'walk-about', where a perusal of the content of their refrigerators was used as a point of departure for a conversation about meal habits. Thematic analysis was undertaken with the reliability and trustworthiness of the analysis assured through verification of codes and discussions among the research team. The analysis provided a nuanced account of the sustainability-related consumption routines, including the intersection and interconnection of meanings, competencies, and material infrastructures (see Appendix B).

Finally, we explored consumer preferences for online food platforms' value propositions and the effect of COVID-19 on platforms' customer base. An online survey was carried out, which drew on the findings from the household study to understand the role of these platforms in supporting sustainability goals. Quantitative analysis was employed to profile respondents based on platform choices, level of use, and motivation (see Appendix C).

Integrating a range of research methodologies, including desk research, website analysis, ethnographic interviews, photo diary, kitchen 'walk-about', and surveys, enabled us to attain a comprehensive understanding of online food provisioning services and sustainable consumption practices. This approach facilitated data triangulation, bolstering the credibility and robustness of our findings. Anchored in social practice theory and sustainability studies, our methodology's adaptability and interdisciplinary perspective ensure that our research is not only rigorous but also universally applicable. In sum, our approach delivers a contextually informed, well-rounded, and credible analysis of the subject matter, significantly enhancing the relevance of our findings for a broader audience.

### *Project Results*

Across the three work packages (WPs), it became evident that food-related motivations and practices among households and individuals exhibited significant variation, mirroring and reshaping the differences observed within the European online food market. In this context, our research uncovers distinct national characteristics. In Germany, a substantial portion of individuals engage in online food purchases, supported by a flourishing landscape of independent stores that prioritise local and sustainable products. In contrast, Ireland records a lower rate of online food buying, with Alternative Food Networks (AFN) taking a prominent role in championing sustainability. Italy, despite a relatively low online shopper base, hosts a vibrant market characterised by a multitude of online providers, especially online-only and independent stores, harmonising with the essence of Italian gastronomic culture. Norway and Sweden both witness the proliferation of online-only shops, yet they diverge in their sustainability commitments, as Norway leans towards local food through platforms and mass retail channels, while Sweden's retail channels emphasize environmental concerns. These findings spotlight the intricate dynamics of the European online food sector, reflecting national economic structures, consumer preferences, and sustainability endeavours.

When examining purchasing patterns, we also found large differences between countries in terms of shopping habits, online purchasing, how products were communicated, and market structures. Our results suggest that while there is no such thing as a 'European consumer', there are some basic commonalities. Although we noticed the existence of a diffused awareness of food sustainability, the extent to which households are moving towards more sustainable consumption behaviours was somewhat limited. Solutions that require alteration to the routines and rhythms of the household's everyday life often meet with some resistance due to the perceived time, effort, and increased knowledge and skill required to make the necessary changes.

Notwithstanding this, there appears to be a willingness to engage, at least to some degree, in environmentally friendly food practices on the part of most people. The sustainability challenge has been more fully embraced by some, both in their product consumption choices and food practices. These 'pioneers' tend to engage often with alternative food networks and consciously select these networks as a means of expressing a sustainable food lifestyle. However, as demonstrated through our household studies, sustainability was not necessarily the main reason for choosing digital AFNs. Rather it was their ability to offer discounts on food, their novelty, or their convenience.

Overall, the findings show that new digital food platforms allow consumers to pursue various sustainability strategies, such as eating more vegetables, buying local food or reducing waste. The choice architecture in the retail space has a considerable impact on household sustainability; the fore-fronting of products and the use of promotional cues in the retail store often result in over-purchasing and waste, while the choice architecture of online food platforms appeared to result in purchasing more based on need and less on impulse. However, the results also show that these new food platforms were not always successful in enabling these forms of sustainable food consumption. The failures of food platforms to expedite sustainable consumption were in turn largely connected to the complexity of food household consumption. Food consumption exists in a nexus of household practices (Fuentes & Samsioe, 2020). Changing how one shops for food, for example, is connected to a series of food (cooking, eating, storing food) and non-food (working, picking up kids, exercising) practices. Change is, therefore, both difficult and can have unforeseen (ripple) effects on other habits. Therefore, while the motivation to be more sustainable can lead consumers to try alternative food supply platforms or purchase more sustainable foods, the household has to transition to new patterns of food behaviour to realise its sustainability goals. This often requires significant 'consumer work', additional resources, and increased knowledge and skills (Fuentes & Samsioe, 2020). One solution will not fit all, and consequently, the platform's role in enabling more sustainable practice can vary considerably.

Moreover, we observed during the research that sustainable provisioning was sometimes a 'by-product' for the consumer—for example, when buying from an online food platform selling local and organic food was primarily motivated by time-saving. From these insights, we developed several recommendations for food suppliers who wish to promote sustainability through digital platforms:

When developing services and products, it is crucial to consider the entire everyday life context in which they exist, rather than focusing solely on individual practices such as purchasing. Consumers desire more user-friendly applications that address their needs around navigation, shopping lists, delivery frequency, and labelling. Food services should thus develop scripts which make it convenient to engage in sustainable food practices.

Many consumers find sustainability to be a confusing concept and often associate local origin with sustainability. We suggest that convenience and origin could serve as significant variables for platform segmentation. However, minimising the emphasis on sustainability could result in unintended consequences, potentially encouraging economic, convenient, or hedonistic shopping habits that overshadow sustainable choices and even lead to excessive consumption. The research drew attention to the potential ways that the platform owner can facilitate and support more sustainable food practices through, for example, creative communication and incentivising more circular food packaging choices. Co-design with consumers is also of particular importance. The intended use of a digital technology for food provisioning does not necessarily correlate with how it is used by consumers at home. Hence, future innovations should be co-designed with consumers to align scripts more closely with their needs and their views of sustainability.

Designing interventions that enable households to better address sustainable food consumption practices offers value at both an individual and a societal level. Insights from this research strongly support the approach of targeting household routines to promote sustainable food consumption in the home. Based on our findings, households' sustainable consumption practices cannot be reduced to consumers alone; rather, enabling sustainable consumption practices must be achieved through reinforcing collective action among all the stakeholders in the food supply chain. Recommendations to guide practical actions are identified, such as promoting the consumption of seasonal foods, nurturing social connection (C2C, B2C, and C2B), improving cooking knowledge and skills and preservation techniques and skills, and addressing packaging challenges. Food manufacturers, retailers, and policymakers need to continue to support product and process developments within the sustainability domain (including packaging and waste minimisation), incentivise more consumption of seasonal foods, promote social connections and knowledge exchange, and additionally consider the sustainability consequences of promotions that are designed around healthy eating and value for money.

## 4 About the Chapters in this Book

Organised in nine chapters with an extensive methodological appendix, the book brings together theoretical and empirical reflections on the role of new technologies and digital platforms in food provision. The book covers several topics based on unique data collected during the PLATEFORMS project. Questioning whether the available alternatives to conventional food purchasing represent a real opportunity to promote more sustainable food systems, the chapters that make up the anthology provide a comprehensive overview and a series of in-depth empirical explorations of how and whether new food supply systems help to structure and organise more sustainable food practices.

Using a unique dataset collected in 2019, Filippo Oncini discusses in his chapter '*Online food provisioning services and where to find them: Pipelines, platforms and the rise of dark store*' some key findings regarding the characteristics of online food procurement services in the five countries studied within the PLATEFORMS project. Building on Parkers et al.'s (2016) distinction between pipelines and platforms, the chapter provides a classification of online providers to highlight similarities and differences among countries. The data collected indicate that platforms tend, on average, to present themselves as more sustainable than pipelines. Yet, while in the face of an unprecedented sustainability crisis, the ability of (some) platforms to reconnect producers and consumers by offering a wide variety of food products might seem attractive, the chapter highlights that, at the moment, these alternatives still seem to involve marginal numbers of consumers.

In the third chapter '*Sustainable and Purchasing Behaviour of Online Food Shoppers: Survey Results from Italy, Ireland, and Germany*', Frode Alfnes, Silvia Biasetton, Margherita Brunori, and Mary McCarthy present the results of a survey on online platform users conducted in January 2021, thus at the height of the lockdown. Examining age and gender differences and purchasing behaviours, the chapter identifies online food consumers in Italy, Germany, and Ireland and discusses the variations in purchasing behaviour and frequency between early adopters and newcomers. As it turns out, most respondents expect to maintain or increase their online grocery purchases in the future, although expectations among newcomers vary, with a higher percentage of Irish novices expecting to decrease their online grocery purchases post-pandemic than German and Italian newcomers.

From analysing the structural characteristics of different online food provisioning services, Margherita Brunori's chapter '*Driving the digital and sustainable transition through law: assessing the food consumer's legal toolkit*' completes this general overview by discussing the regulatory and policy framework surrounding digital food platforms. She examines how the sustainability ambitions of the EU's Green Deal challenge and stretch consumer rights and protection. The chapter adopts a food consumer-centred perspective and interrogates the EU's legal framework to clarify what tools are granted by legislatures [i.e. national Parliaments] to citizens to fulfil their food rights, needs, and choices while navigating the digital marketplace. The study gives an account of the most recent advances in EU food law of so-called digital consumer rights while offering a concise assessment of the current challenges and future directions of the legal environment in food.

The chapter '*Infrastructure, impulsivity, and waste. Exploring the (un) sustainable routines of mainstream food shoppers*' by Claire O'Neill, Shadi Hashem, Mary McCarthy, and Christopher Moran offer an in-depth look at the purchasing routines of so-called mainstream consumers and uses a social practices framework to capture the purchasing habits of 21 Irish shoppers. The analysis focuses on how purchasing practices contribute to food waste, distinguishing between organised and disorganised consumers. The results show that infrastructure and impulse-buying both have implications for food waste at the household level. Online technologies, such as supermarket websites and apps, appear to help reorganise the shopping practices of disorganised households. However, due to the material infrastructure of the online platform, shoppers who purchase online tend to repeat orders without checking stock, producing a surplus of food, especially healthy food. In other words, although successful in organising the weekly shopping experience and ensuring adequate food stocks and meal planning, the online infrastructure may facilitate the perpetuation of healthy food being thrown away, week after week.

The theme of food waste is also at the centre of the chapter '*Making the Consumption of Food Circular: The Karma App and the Re-qualifications of Surplus Food*' by Christian Fuentes, Emma Samsioe, Maria Fuentes, and Jenny Ivre. The chapter investigates the Swedish smartphone application 'Karma' whose aim is to reduce food waste by allowing restaurants and retailers to sell unsold food surpluses to consumers at discounted prices. Through a series of digital observations to understand how the app works and by using 10 ethnographic interviews with Karma users, the authors discern several cases where the app is valued not primarily for its sustainability benefits but rather for its ability to offer discounts on food, either because of its novelty aspects or the convenience it offers. From a critical point of view, the analysis shows that the resulting food consumption pattern is circular, but only sometimes sustainable.

The article '*From grassroots to platforms: how digitalisation reconfigures learning and engagement with food*' by Francesca Forno, Suse Brettin, Christopher Moran, and Sandra Cajic moves the focus to those models of distribution and consumption that enable direct contact between producers and consumers, such as alternative food networks (AFN), both physical and digital. The chapter questions how digitalisation reconfigures the 'learning by doing' which has often been highlighted as an important feature of alternative food provisioning systems. As underlined by many studies, through alternative provisioning networks, people do not just find the food they want to buy. The ongoing networking activities facilitate different learning processes by harnessing individuals' capacity to understand the intricate interdependence of environmental, social, and economic issues. By starting from the assumption that digitalisation represents a new form of intermediation and using 83 in-depth interviews conducted in Ireland, Germany, and Italy, this chapter seeks to compare and contrast consumer learning and engagement processes in physical and digital AFNs.

Maintaining the focus on people who buy through alternative provisioning channels, Alice Dal Gobbo's chapter '*Food, health, and sustainability: choice, care, alternatives*' investigates how people in these networks articulate the health-sustainability nexus in their daily food practices. While for almost all of the 45 people interviewed in Milan, the main reason for approaching alternative provisioning channels was to provide themselves and their families with good-tasting food redolent of ideas of naturalness, artisanal production, small-scale and humane agriculture, and even spirituality, informants tended to 'assemble' questions related to health and sustainability in different ways. The most transformative articulation was found in the context of collective consumption experiences, where bodily health concerns are understood as directly related to community and territorial health.

The new demands of food-related care work not only follow different logics but also affect women in particular. Focusing their attention on two urban contexts—Berlin and Milan—and analysing the evidence from 85 in-depth interviews, Suse Brettin, Meike Bruchner, and Natalia Magnani in the chapter '*A plate full of food and care. Investigating care practices and gender relations at the crossroads of sustainability, innovative practices, and digital platforms*' observe how it is mostly women who are reinventing their activities in the kitchen by using digital technologies as a way of not feeling overwhelmed by demands of sustainability. Rather than an instrumental motivation to save time and money, obtaining healthy and sustainable food through technology seems to offer support and enable care work towards family and household members and other actors in the food chain, small-scale farmers especially.

Given the dynamics and diversity of OFPS, Jiaxi Chen's concluding chapter '*Online Food Provisioning Services (OFPS) as Sociotechnical Assemblages: Three Theoretical Threads and a Proposition for Further Research*' argues that it is important to find theoretical approaches to conceptualise and understand how these OFPS interact with consumption practices in the context of digitisation and the sustainable transition. The chapter compares the similarities and differences between three theoretical strands often used in studies of science, technology, and society—Actor-Network Theory (ANT), Theories of Practice (ToP), and the Multilevel Perspective (MLP)—and highlights how the vision of socio-technical assemblage shared by these three theoretical approaches can be useful in generating an in-depth understanding of OFPS-based food practices.

The book includes a detailed methodological appendix which describes the three data collection techniques used. This additional part aims to allow and stimulate the possibility of updating, replicating, and validating the results obtained through the PLATEFORMS study. Given the recent changes in the sector, subsequent studies could deepen our understanding of the phenomenon and allow the processes at work to be read more accurately and consciously.

The focus of the book is very much on food. There is, however, an opportunity to reflect more broadly on what the key insights say about consumption and everyday life more generally. Sociological studies of consumption have shown that the consumption practices of households differ significantly on the basis of socio-economic and socio-cultural characteristics such as household type, income, race/ethnicity, age, sexuality, gender composition, and social class. Consumption inequality is also mirrored in digital access, with a digital divide existing between low-, medium-, and high-income households, even in developed economies. So, one question would be whether digital technologies increase the division of healthy and sustainable eating that already exists between high- and low-income households.

Finally, while much-existing research examines how digital technologies can support more modes of consumption that are assumed to come with environmental benefits, these studies seldom quantify the environmental effects of the new modes of practices enabled by these digital technologies. There is a need to explore how food provision practices are interlinked with other household practices, such as transport, and whether these new modes of consumption require a higher level of mobility.

## References

Alfredsson, E., Bengtsson, M., Brown, H., Isenhour, C., Lorek, S., Stevis, D., & Vergragt, P. (2018). Why achieving the Paris Agreement requires reduced overall consumption and production, Sustainability. *Science, Practice and Policy, 14*(1), 1–5. https://doi.org/10.1080/15487733.2018.1458815

Bos, E., & Owen, L. (2016). Virtual reconnection: The online spaces of alternative food networks in England. *Journal of Rural Studies, 45*, 1–14. https://doi.org/10.1016/j.jrurstud.2016.02.016

Cooper, G., Green, N., Burningham, K., Evans, D., & Jackson, T. (2012). Unravelling the threads: Discourses of sustainability and consumption in an online forum. *Environmental Communication, 6*(1), 101–118. https://doi.org/10.1080/17524032.2011.642080

Corsi, A., & Novelli, S. (2018). Determinants of participation in AFNs and its value for consumers. In A. Corsi, F. Barbera, E. Dansero, & C. Peano (Eds.), *Alternative Food Networks* (pp. 57–86). Springer International Publishing.

Dal Gobbo, A., Forno, F., & Magnani, M. (2022). Making good food more practicable? The reconfiguration of alternative food provisioning in the online world. *Sustainable Production and Consumption, 29*, 862–871. https://doi.org/10.1016/j.spc.2021.07.023

Davies, A. R., & Legg, R. (2018). Fare sharing: interrogating the nexus of ICT, urban food sharing, and sustainability. *Food Culture & Society, 8014*, 1–22. https://doi.org/10.1080/15528014.2018.1427924

De Bernardi, P., Bertello, A., & Venuti, F. (2019). Online and on-site interactions within alternative food networks: Sustainability impact of knowledge-sharing practices. *Sustainability (Switzerland), 11*(5), 1457. https://doi.org/10.3390/su11051457

Dulsrud, A., & Bygstad, B. (2022). Digital ecosystems as social institutions: Exploring the role of consumption through four research streams of digital ecosystems. *Consumption and Society, 1*(1), 99–119. https://doi.org/10.1332/QUPH6141

Forssell, S., & Lankoski, L. (2015). The sustainability promise of alternative food networks: An examination through 'alternative' characteristics. *Agriculture and Human Values, 32*(1), 63–75. https://doi.org/10.1007/s10460-014-9516-4

Fuentes, C., Cegrell, O., & Vesterinen, J. (2021). Digitally enabling sustainable food shopping: App glitches, practice conflicts, and digital failure. *Journal of Retailing and Consumer Services, 61*. https://doi.org/10.1016/j.jretconser.2021.102546

Fuentes, C., & Fuentes, M. (2022). Infrastructuring alternative markets: Enabling the exchange of local food through patchworking. *Journal of Rural Studies, 97*, 13–22. https://doi.org/10.1016/j.jrurstud.2022.05.022

Fuentes, C., & Samsioe, E. (2020). Devising food consumption: Complex households and the socio-material work of meal box schemes. *Consumption Markets & Culture, 24*(5), 492–511. https://doi.org/10.1080/10253866.2020.1810027

Fuentes, C., & Sörum, N. (2019). Agencing ethical consumers: Smartphone apps and the socio-material reconfiguration of everyday life. *Consumption, Markets and Culture, 22*(2), 131–156. https://doi.org/10.1080/10253866.2018.1456428

Geels, F. W., McMeekin, A., Mylan, J., & Southerton, D. (2015). A critical appraisal of sustainable consumption and production research: The reformist, revolutionary and reconfiguration positions. *Global Environmental Change, 34*, 1–12. https://doi.org/10.1016/j.gloenvcha.2015.04.013

Goodman, D., DuPuis, E. M., & Goodman, M. K. (2012). *Alternative food networks: Knowledge, practice, and politics*. Routledge.

Graham, M., & Haarstad, H. (2011). Transparency and development: Ethical consumption through Web 2.0 and the Internet of things. *Information Technologies & International Development, 7*(1), 1–18.

Gustavsson, J., Cederberg, C., Sonesson, U., Van Otterdijk, R., & Meybeck, A. (2011). *Global food losses and food waste*. Food and Agriculture Organization of the United Nations (FAO).

Hertz, F. D., & Halkier, B. (2017). Meal box schemes a convenient way to avoid convenience food? Uses and understandings of meal box schemes among Danish consumers. *Appetite, 114*, 232–239. https://doi.org/10.1016/j.appet.2017.03.016

Holmberg, U., Hansson L., & Post A. (2022). Reorganizing grocery shopping practices—The case of elderly consumers. International Review of Retail, Distribution and Consumer Research (forthcoming). https://doi.org/10.1080/09593969.2022.2085137

Huyghe, E., Verstraeten, J., Geuens, M., & Van Kerckhove, A. (2017). Clicks as a healthy alternative to bricks: How online grocery shopping reduces vice purchases. *Journal of Marketing Research, 54*(1), 61–74. https://doi.org/10.1509/jmr.14.049

IPCC. (2019). IPCC special report on climate change, desertification, land degradation, sustainable land management, food security, and Greenhouse gas fluxes in Terrestrial Ecosystems. Approved Draft. 07 August 2019.

Michelini, L., Principato, L., & Iasevoli, G. (2018). Understanding food sharing models to tackle sustainability challenges. *Ecological Economics, 145*, 205–217. https://doi.org/10.1016/j.ecolecon.2017.09.009
Nemes, G., Chiffoleau, Y., Zollet, S., Collison, M., Benedek, Z., Colantuono, F., Dulsrud, A., Fiore, M., Holtkamp, C., Kim, T.-Y., Korzun, M., Mesa-Manzano, R., Reckinger, R., Ruiz-Martínez, I., Smith, K., Tamura, N., Viteri, M. L., & Orbán, É. (2021). The impact of COVID-19 on alternative and local food systems and the potential for the sustainability transition: Insights from 13 countries. *Sustainable Production and Consumption, 28*, 591–599. https://doi.org/10.1016/j.spc.2021.06.022
Oncini, F., Bozzini, E., Forno, F., & Magnani, N. (2020). Towards food platforms? An analysis of online food provisioning services in Italy. *Geoforum, 114*, 172–180. https://doi.org/10.1016/j.geoforum.2020.06.004
Parker, G., Van Alstyne, M., & Choudary, S. P. (2016). Platform revolution: How networked markets are transforming the economy and how to make them work for you (1st ed). W. W. Norton & Company.
Samsioe, E., & Fuentes, C. (2021). Digitalizing shopping routines: Re-organizing household practices to enable sustainable food provisioning. *Sustainable Production and Consumption, 29*, 807–819. https://doi.org/10.1016/j.spc.2021.07.019
Schneider, T., Eli, K., Dolan, C., & Ulijaszek, S. (Eds.). (2018). *Digital food activism*. Routledge.
Seyfang, G. (2005). Shopping for sustainability: Can sustainable consumption promote ecological citizenship? *Environmental Politics, 14*(2), 290–306. https://doi.org/10.1080/09644010500055209
Sonnino, R., & Marsden, T. (2006). Beyond the divide: Rethinking relationships between alternative and conventional food networks in Europe. *Journal of Economic Geography, 6*(2), 181–199. https://doi.org/10.1093/jeg/lbi006
Southerton, D., & Fuentes, C. (2022). Digital platforms and sustainable food consumption transitions. *Sustainable Production and Consumption, 29*, 805–806. https://doi.org/10.1016/j.spc.2021.08.023
Srnicek, N. (2017). *Platform capitalism*. Polity Press.
Zuboff, S. (2019). *The age of surveillance capitalism: The fight for a human future at the new frontier of power* (1st ed.). Profile Books.

# Online Food Provisioning Services and Where to Find Them: Pipelines, Platforms and the Rise of Dark Stores

*Filippo Oncini*

## 1 Introduction

The past decade has been characterised by the success and expansion of platform-based companies. Such paradigm shift reconfigured work relationships and labour market dynamics worldwide, creating very diverse conditions of employment depending on workers' skills, locations and the platform's level of control (Forde et al., 2017; Vallas & Schor, 2020). The rapid emergence of the platform economy, and its capacity to operate across very different sectors and geographical contexts, is the cause of current ambiguities over its intermediatory role. Opportunities for flexible work hours and autonomy are counterbalanced by insecure forms of employment, weak employment rights, new forms of discrimination and

F. Oncini (✉)
Maastricht Sustainability Institute, Maastricht University,
Maastricht, The Netherlands
e-mail: filippo.oncini@maastrichtuniversity.nl



A. Dulsrud, F. Forno (eds.), *Digital Food Provisioning in Times of Multiple Crises*, Consumption and Public Life,
https://doi.org/10.1007/978-3-031-46323-5_2

obstacles to implementing effective regulation (Forde et al., 2017; van Dijck, 2021; Yates, 2020; see also Brunori's chapter in this collection). Ease of access to platform work implies that participants can differ greatly in motivations and levels of dependence, resulting in greater fragmentation in the political orientation and demand for regulation (Zuboff, 2019). In addition, the working conditions and earning prospects change depending on the platform sector: clearly, cloud-based consultants' economic conditions and precarity differ considerably from those of drivers and cleaners on Uber and Helpling, despite both taking part in platform-mediated work. Finally, critical theorists cast a shadow over the enormous control that platform companies have over personal data circulation and exploitation, pointing out how privately owned, predictive algorithms have given rise to a new era based on surveillance and manipulation (Zuboff, 2019). The increasing complexity of the platform-based economy caused some authors to refer to it as digital or platform *ecosystems*, a term that permits taking into account the open, adaptive, sociotechnical and political nature of a digital network connecting people and organisations (Dulsrud & Bygstad, 2022; Yates, 2020; van Dijck, 2021).

Roughly over the same period, online shopping for groceries and meals started attracting the attention of customers throughout Europe. Despite striking differences between countries, in the EU the percentage of people that purchased food online at least once over the past year increased from 4% in 2010 to 15% in 2019 (Eurostat, 2021). In the context of the Covid-19 pandemic, reliance on online food shopping further increased, partly accelerating preexisting trends. For instance, a recent cross-country comparison of Covid-19 responses by alternative and local food systems (ALFS) illustrated how the crisis pushed many organisations to adopt technological and social innovations, for instance, digitalising physical markets or the provision logistics (Nemes et al., 2021). Despite the growth of food e-commerce however, an actual platform revolution for *grocery* shopping is still far from taking place (Oncini et al., 2020). On the one side, as the Eurostat (2021) data show, if we consider the purchase of just food or beverages from stores or from meal-kit providers over the previous three months, the EU-27 average is only 10%. To put it into perspective, this is roughly the same percentage of people that rented an accommodation online, with the difference that food consumption occurs much more frequently and ubiquitously than does overnighting away from home. On the other side, online food provisioning services (OFPS) are not necessarily based on the platform model, but rather present a constellation of several types of online shops (Oncini et al., 2020) and therefore can be seen as a digital ecosystem (Dulsrud & Bygstad, 2022).

In this regard, the distinction between pipeline and platform advanced by Parker et al. (2016) can be useful to draw the line and separate two main macro-models. According to the authors, the first identifies a 'business that employs a step-by-step arrangement for creating and transferring value, with producers at one end and consumers at the other' (Parker et al., 2016, p. 6). Platforms are instead 'based on enabling value creating interactions between external producers and consumers. The platform provides an open, participative infrastructure for these interactions and sets governance conditions for them. The platform's overarching purpose: to consummate matches among users and facilitate the exchange of goods, services, or social currency, thereby enabling value creation for all participants' (Parker et al., 2016, p. 5).

Among OFPS, pipelines are single-sided online food shops where the owner of the service controls the selection, presentation and shipment or collection of the groceries (e.g., the online shop of a brick-and-mortar delicatessen store). Food platforms are instead two-sided online food markets where the intermediary facilitates the match between food producers or retailers and consumers (B2C), but also between consumers and consumers (C2C). As in the case of Airbnb or Uber in the accommodation and mobility services, respectively, the food platform offers a technostructure that permits the match between two sides that aim to conclude a transaction (Oncini et al., 2020).

Despite an abundance of literature on the platform economy and on online grocery shopping (e.g., Piroth et al., 2020; Samsioe & Fuentes, 2022; Vallas & Schor, 2020), research to date has not yet investigated the differences between pipeline and platform models, especially from a cross-country perspective (but see Nemes et al., 2021). Aiming to fill this gap, this chapter summarises some key findings that emerged from a cross-country study conducted in five European countries (Germany, Ireland, Italy, Norway and Sweden). The chapter is structured as follows. After presenting the data collection strategy and the methods used in the paper, I propose a classification of different pipeline and platform models that could be identified in each country and illustrate their differences with examples from each country. I therefore present some differences that emerged across countries between the models by focusing on (i) the market scope of the models and the geographical scope of the producers they rely on; (ii) the value commitments and (iii) the range of food products sold by each model. I hence move on to analyse the on-demand grocery delivery companies that emerged during the Covid-19 crisis and reflect on their capacity to leverage the most profitable practices of *both* pipelines and platforms. In the conclusions, I discuss new directions for research on online food providers and highlight new possibilities to update and extend the data set.

## 2 Data and Methods

The PLATEFORM data set has been constructed by gathering data on the population of OFPS active in Germany, Ireland, Italy, Norway and Sweden between December 2018 and April 2019. To gather information on all active OFPS, one researcher in each country searched the following keywords, translated into each country's language, on Google.com, Google Play Store and Apple Marketplace search engines: 'online food shopping', 'online food purchase', 'online grocery shopping', 'organic food online', 'local food online', 'food sharing apps'. The search was subsequently refined adding NUTS2 regional codes to each keyword to increase the probability of finding smaller or local provisioning services. Because the project aims to map the OFPS that potentially allow people to organise the grocery shopping online, we excluded all the OFPS selling either (i) only one type of food category (e.g., only meat, only dairy, only vegetables, only wine) or (ii) from only one brand (e.g., only Barilla), or (iii) from only one producer. After the initial screening and deletion of duplicates, 524 websites were identified as OFPS across all countries. The OFPS are distributed as follows: 86 in Germany, 90 in Ireland, 211 in Italy, 81 in Norway and 56 in Sweden.

The content analysis of the websites followed a two-stage process. A first version of the codebook was used to conduct the analysis on all the retrieved OFPS. The codebook contained indications on how to operationalise information on the following areas of concern from each website: organisational form of the e-commerce, food categories, sales mechanisms, digital presence, and value commitments. The codebook (available upon request) contains full information on all variables, the coding procedure and the specific information gathered by each coder. Therefore, all researchers involved in the project gathered in an online meeting to discuss the most ambiguous variables. These variables were then reanalysed on a separate data set containing a random selection of 30 OFPS based in England. This choice allowed coders to analyse the reliability data set in the only language that was comprehensible to all. Following the results of the reliability test, the average pairwise percent agreement, Fleiss' K and Krippendorff's Alpha have been used to discard or simplify (i.e., reducing the number of response categories) the less reliable variables.

To highlight some of the most important findings, here I use descriptive statistics of the following variables: year of the foundation of the online provider, type of online food provider (mass retail, independent store, alternative food network, online-only, platform), a dummy indicating whether the provider has a local market scope (ships products only locally) and relies on local producers, a dummy indicating whether commitments towards local foods are reported, the total number of commitments reported (local food, waste reduction, environment, equity, nutrient needs)[1] and the total number of food categories sold (bakery, cereals, meat, fish, fruit, vegetables, alcoholic drinks, nonalcoholic drinks, preserves, snacks, cured meat, dairy, ready-made).

## 3 Categorising Online Food Providers

Figure 1 uses the pooled sample to illustrate the number of providers by year of foundation. In line with data indicating the growing success of food e-commerce all over Europe, the number of new online services selling groceries has almost steadily increased from 1998 to 2018, apart from a small decline in 2017 and 2018. Although these providers are from the same sector of the economy, their respective business models can change considerably. For this reason, I distinguish between five different models: mass retail channel, independent stores, online-only stores, alternative food networks (AFN hereafter) and platforms. The first four can all be considered pipelines. In other words, these shops do not work as digital

[1] Commitments were obtained from the OFPS websites by means of content analysis. We created a dummy variable distinguishing whether the website reported clear statements on the products' origins (whether they come from a specific region or area), on waste reduction (whether the organisation tries to avoid or reduce food waste), on environmental protection (generic statements on sustainability), on equity/fair trade (whether the organisation is attentive to the working conditions of their employees and/or of the producers), and whether the organisation clearly targets at least one group having special dietary requirements (e.g., celiac, vegan, lactose-intolerant, etc.). The total number of commitments is the simple sum of the five dummies which was then used to obtain the mean across OFPS typologies.

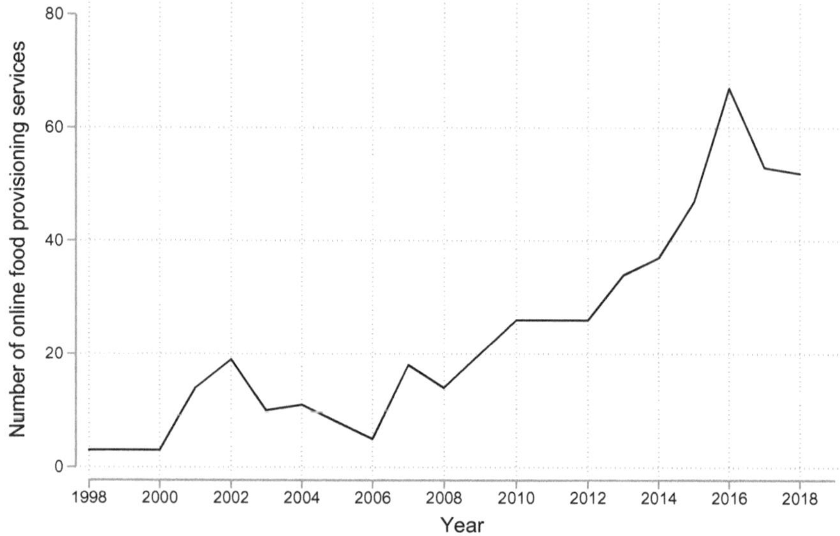


**Fig. 1** Total number of OFPS by year of foundation. ($N$ = 499 (25 "date not reported"))

intermediators facilitating a match between consumers (C2C), between consumers and producers, or between consumers and retailers (B2C).

The mass retail channel category (9.4% of the total) comprises all those supermarket chains that allow customers to purchase products online through their websites. Independent stores (31.5%) are instead the e-commerce websites of many small food boutiques and grocers, usually specialised in a few product categories. AFN (6.9%) generally refers to organisations that propose alternative forms of food production and distribution. These are conceived as being in opposition to conventional food chains (Forno & Graziano, 2014; see also Forno et al.'s chapter in this collection). Online-only stores (42.4%) differ from the other three pipeline OFPS because they exist only in their digital version. Finally, platforms represent 9.9% of the sample.

Table 1 illustrates for each country an example of OFPS divided by typology. Mass retail includes some of the biggest supermarket chains that also operate online, such as Rewe, Tesco, Esselunga, Coop Norway, and Willys. These chains have been offering the option to receive groceries at

**Table 1** Examples of OFPS by typology and country

| | *Mass retail* | *Independent store* | *Online-only pipeline* | *AFN* | *Platform* |
|---|---|---|---|---|---|
| Germany | *Rewe* | *Flotte Karotte* | *Hello Fresh* | *Landkorb* | *Food Sharing* |
| Ireland | *Tesco* | *On the pig's back* | *Irish Gourmet* | *Organic Republic* | *Buymie* |
| Italy | *Esselunga* | *Lunelli* | *inSicilia* | *Gasgos* | *Food Genuine* |
| Norway | *Coop Norge* | *Etikken* | *Nettmat* | *Fritt og vilt* | *Dyrket* |
| Sweden | *Willys* | *Happy Food Store* | *Norrbete* | – | *Matspar* |

home or to pick them up at the store for several years now. Independent stores are instead relatively small food shops or delicatessens based in specific locations. These shops can be very specialised; for instance, *On the Pig's Back* is a small charcuterie shop based in Cork (Ireland) that mainly focuses on meats and cheeses; *Flotte Karotte* instead sells organic-only products, but has a wider range of food categories. Differently however, *Lunelli* is a small independent shop with two stores, one in the city of Trento and one in the town of Sarche, selling online a wide variety of typical food produced in Trentino (Italy). Online-only stores can be extremely heterogeneous, because they range from shops selling US confectionary products to food boxes with precise ingredients to prepare a meal (e.g., *Hello Fresh* and *Nettmat*). In Italy, many of them sell typical regional products, as is the case of *inSicilia*, an e-commerce specialised in Sicilian products. AFN are instead all characterised by attempts to shorten distances between producers and consumers while committing to the social, economic and environmental dimensions of sustainable food production, distribution and consumption. *Gasgos* is an interesting example, because the e-commerce is managed by a solidarity purchasing group and allows only members of the association to shop for groceries online and pick them up in specific locations. Finally, the five platforms selected illustrate both B2C and C2C business models. *Food Sharing* is a German-based C2C platform that allows consumers to share purchased food that otherwise would go to waste; *Buymie* and *Matspar* are instead B2C platforms based in Ireland and Norway, respectively, and allow consumers to order groceries online from supermarkets and receive them directly at home. *Food Genuine* and *Dyrket* are also B2C platforms, but in both cases, consumers can shop directly from a list of selected producers who oversee shipping the products.

Finally Figure 2 and Table 2 below show that country differences are very marked, in both relative and absolute terms. In fact, both the total number of online food providers and their distribution present an almost unique configuration (see also Box 1 below). Ireland, for instance, has the lowest number of mass retail channels active online, and the highest proportion of AFN (22.2%). Independent stores are the leading typology in

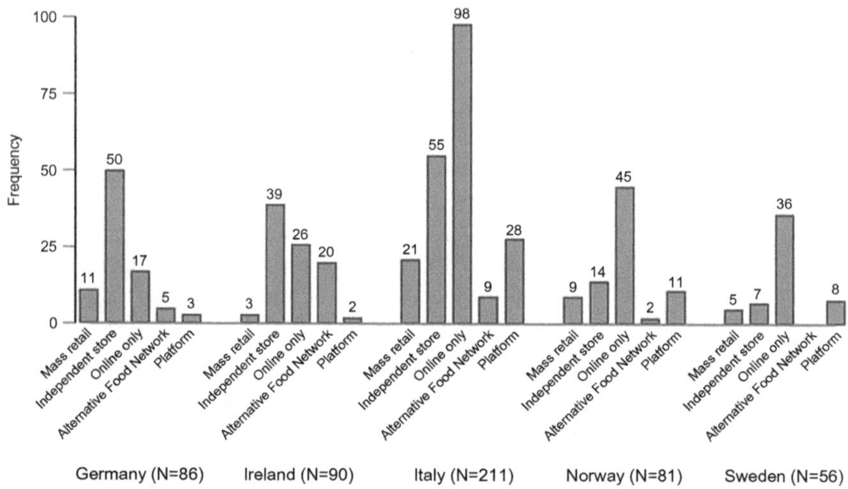


**Fig. 2** Total number of OFPS by typology and country. ($N = 524$)

**Table 2** Types of OFPS by country. ($N = 524$)

| *Country* | *Mass retail channel* | *Independent store* | *Online-only* | *AFN* | *Platforms* |
|---|---|---|---|---|---|
| Germany | 12.8 | 58.1 | 19.8 | 5.8 | 3.5 |
| Ireland | 3.3 | 43.3 | 28.9 | 22.2 | 2.2 |
| Italy | 10.0 | 26.1 | 46.5 | 4.3 | 13.3 |
| Norway | 11.1 | 17.3 | 55.6 | 2.5 | 13.6 |
| Sweden | 8.9 | 12.5 | 64.3 | 0.0 | 14.3 |
| Total | 9.4 | 31.5 | 42.4 | 6.9 | 9.9 |

Germany (58.1%) and Ireland (43.3%), whereas they are less common in Scandinavian countries. Online-only stores are instead very active in Sweden (64.3%), Norway (55.6%) and Italy (46.5%), but they are much less common in Germany and Ireland. In absolute terms, Italy has the largest number of providers (211) and specifically of platforms, followed by Ireland (90), Germany (86), Norway (81) and Sweden (56).

## 4 Main Differences Between Models

To analyse the main differences between the five models, in what follows I will concentrate on three pieces of information that permit one to understand the range of action, the value commitments and the number of food items sold. The range of action captures an actual aspect of sustainability that is related to the length of the food supply chain, because local food systems that connect producers and consumers have the potential to be more sustainable. The analysis of a business's value commitments provides an insight into its self-perceived sustainability, whereas the average number of food items sold reflects the comprehensiveness of its food offerings—and potentially its capacity to mimic the more traditional grocery shopping experiences in supermarkets.

To identify the range of action by typology, I focus simultaneously on the producers' geographical scope (if producers can be linked to one specific place) and the market's scope (i.e., where the company ships the food products). Figure 3 simply distinguishes, by country, the percentage of stores that have local geographical scopes on both the production side and the market side. In other words, the light grey bar indicates, for each store typology, the proportion of those services that tend to favour the match between local producers and local consumers, thus potentially being a more environmentally sustainable alternative to the other options. The first thing to notice is that, despite country differences, mass retailers, independent stores and online-only stores do not favour a scope that consists of both local producers and local shipments simultaneously. This is not surprising: mass retailers may have a local market scope because only people within a certain radius of the store or warehouse can access the service; yet their reliance on mass production inevitably implies that the range of food products and producers will be extremely heterogeneous. Independent stores, on the contrary, often sell food from local producers, but in addition they are willing to ship their products nationally or internationally to increase their customer base. Online-only stores can be very

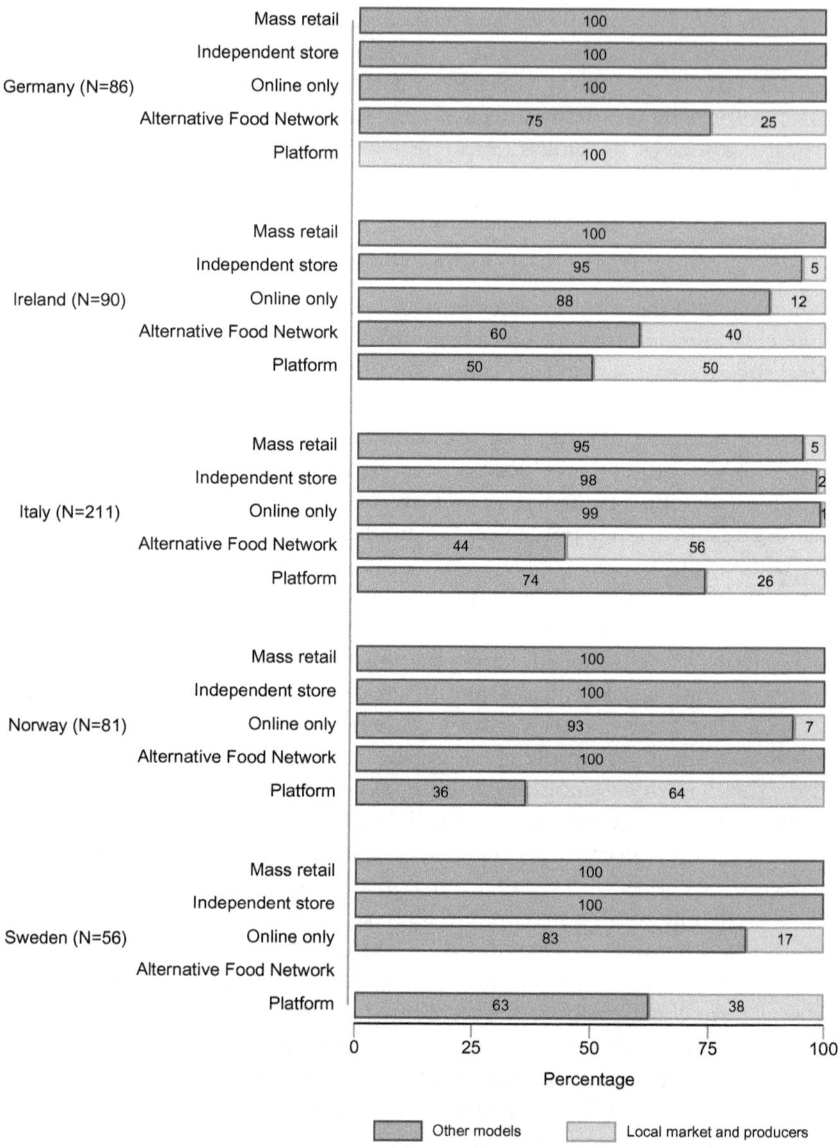


**Fig. 3** Simultaneous presence of a local and geographical market scope by country and type of OFPS

heterogeneous: in Italy, for instance, many of them sell regional food produced locally (Oncini et al., 2020), but similar to independent stores, they are willing to ship their food wherever they can. Differently, we find the highest proportion of localised companies on both the production side and the market side reside among AFN and platforms. Although they rarely tend to be the majority within one provision type, this result suggests that the two models may present some similarities, inasmuch as they tend to see the local dimension as a strength. This local dimension, however, does not imply that the model cannot scale up: the platform known as The Food Assembly, for instance, operates across 9 countries but requires producers and consumers to reside in a 60 km radius. Even so, it should be underlined that even among platforms, the absolute numbers of businesses with local scopes on both the production and the market sides is overall negligible.

Data on value commitments partly reflect the relatively higher importance of the local dimensions for both AFN and platforms. Figure 4 illustrates the percentage of companies that mention a commitment to local food on their websites, broken down by country and business model. In almost all countries, these two models tend to report more often that local food is an important part of their business activities. However, two interesting deviations are worthy of noting. In Germany, independent stores, and not platforms, report the highest presence of mentions concerning local food (86.0%) together with AFN. The only three food platforms active in Germany never report commitments about local food. In Ireland, although platforms and AFN have high percentages of mentions, the online mass retail typology is the most attentive to local food commitments.

Figure 5 illustrates the average number of commitments reported by each typology in each country, reflecting their self-representation in terms of sustainability. For each country, the online food provision models are sorted from the highest to the lowest value in the index. This measure gives in a glance the models that represent themselves as being more sustainable from both an environmental and a social perspective. Four considerations are in order: first, Germany confirms the peculiarity of the online independent stores: on average, they report more commitments. Second, AFN are the models with the highest number of mentions in both Italy and Ireland, whereas in the two Scandinavian countries, the online mass retail channel is the one reporting the highest number of commitments. In Ireland, Italy and Germany however, the situation is different; the mass retail type is always placed third, with an average of around two commitments. Third, online-only pipelines tend to report a low number of commitments in every country. Fourth, platforms have the

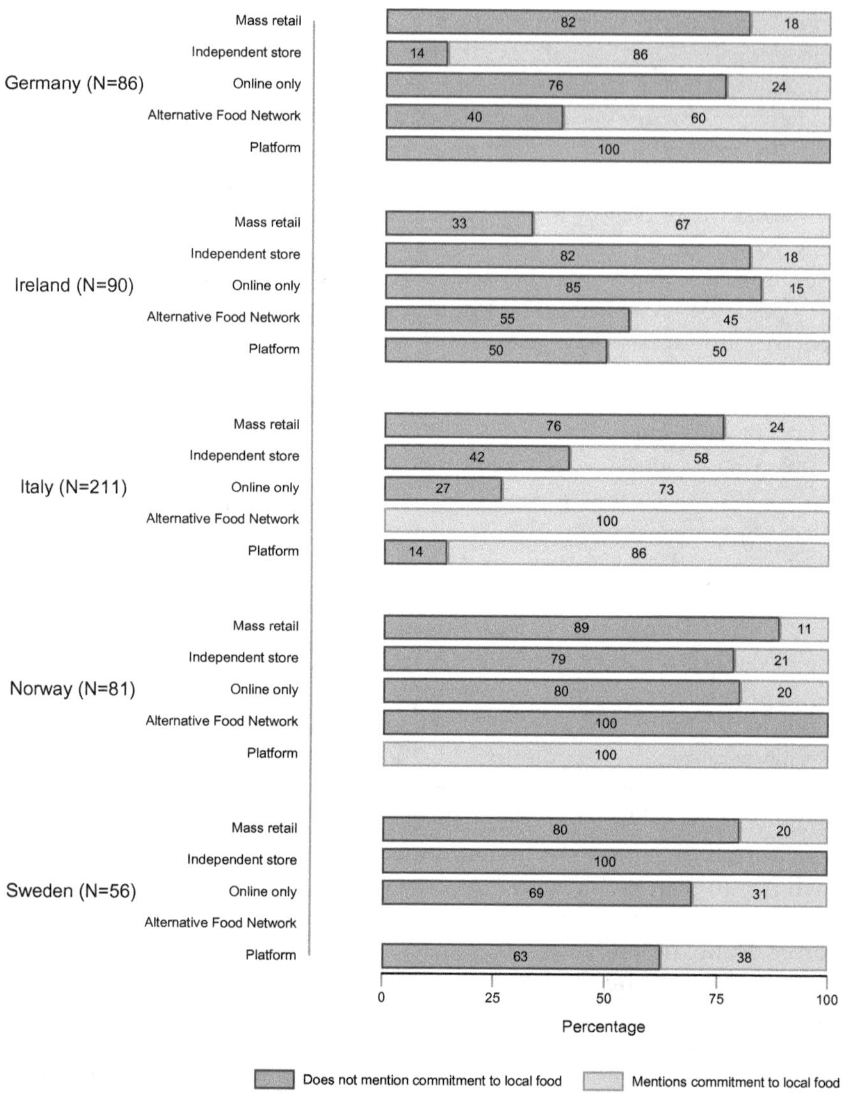


**Fig. 4** Websites reporting commitment to local food by country and type of OFPS

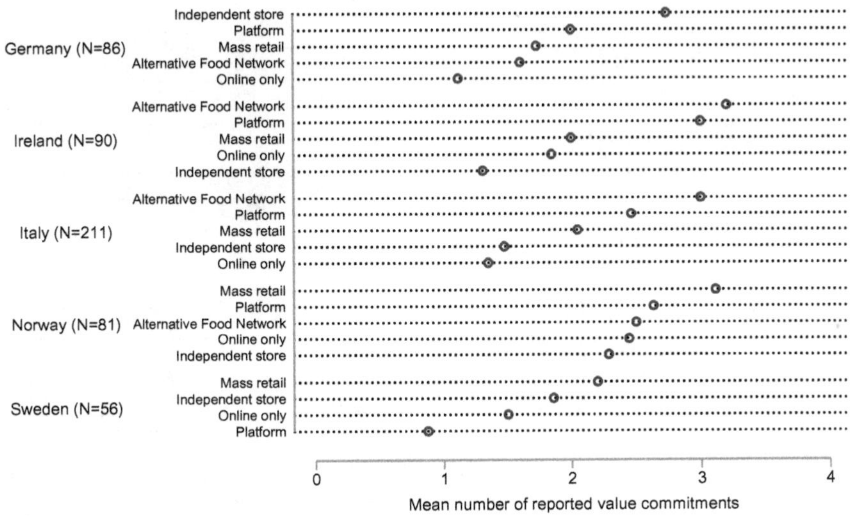


**Fig. 5** Mean number of reported value commitments by country and type of OFPS

second-highest number of commitments in all countries except in Sweden, where they have the lowest average score.[2]

Finally, Figure 6 illustrates the average number of food categories sold within each typology in each country, highlighting the potential for these platforms to serve as comprehensive grocery shopping destination. This is a particularly important dimension, because the heterogeneity of the available food could in principle guarantee that all the grocery shopping could take place in a single online shop because it already happens for in-person

[2]A closer look at the types of commitments also reveals commonalities and differences reported by the online providers in each country (analyses available upon request). Attention to local food and nutritional needs is most prevalent among Italian (67.3% for both commitments) and German (60.7% for both commitments) providers, though local food is also highlighted in the other countries by roughly a quarter of the analysed websites. Attention to the environment is a commitment particularly relevant to providers in Germany, Sweden and Norway (reported by almost half the websites), potentially reflecting that ideas of environmental sustainability are more diffused among the population. Food waste and equity are instead the commitments least reported in all countries, with the exception of Norwegian providers mentioning the latter in 40% of the analysed websites.

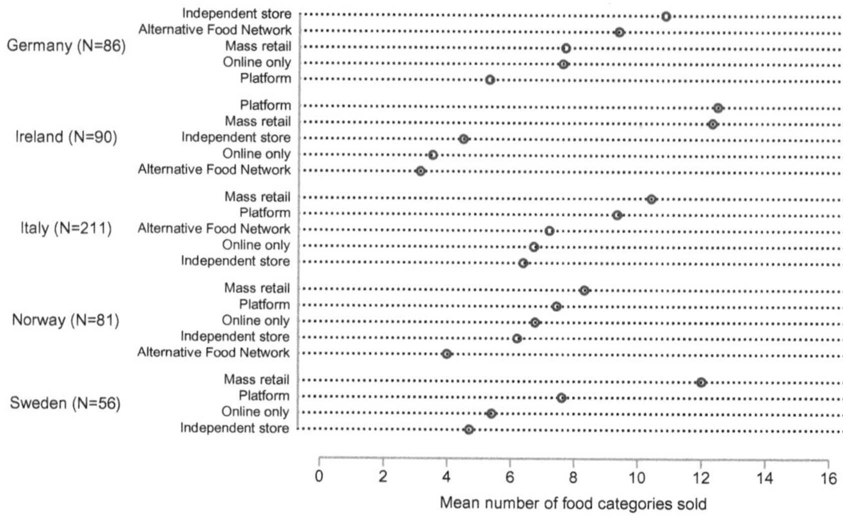


**Fig. 6** Mean number of food categories sold by country and type of OFPS

grocery shopping. Apart from Germany, where online independent food stores have the highest number of food categories sold, the platform and the mass retail models are the ones with the widest range of food categories across all countries.[3] The unique case of German online independent stores stands out and merits further investigation. In any case, the picture suggests that platforms and the mass retail channel are the only models

[3] Additional analyses were conducted on the full sample using latent class analysis (e.g., Oncini & Triventi, 2021) on the 13 food categories and on the 6 identified food clusters: 'Veg and Fruit' is the smallest class (5.2%) and comprises fresh and agricultural products; 'Non-fresh' (30.5%) includes the sale of long-keeping packed products but not alcoholic drinks; 'Ready-made' (9.0%) takes its name from the only item that has a higher probability of endorsing the class, compared to all the other classes; 'Preserves' (23.3%) differs from the 'Non-Fresh' class because it contains long-keeping packed products, cured meat, dairy, and alcoholic drinks. The class named 'Fresh' (7.3%) could instead be considered as an extension of the first one, because it contains, besides fruit and vegetables, all sorts of fresh products with a short expiration date; finally, the class 'All products' (24.8%) contains the services that have the most heterogeneous provisioning of food products, because all items have a high probability of endorsing the class. The 'All products' class, in line with the results shown above, is significantly more present in the mass retail and in the platform models. Results are available upon request.

capable of offering a wide variety of foods. This could be the case for two different reasons shared by all the countries studied: on the one side, the mass retail channel can guarantee the widest offering because all food typologies are already stored in the brick-and-mortar supermarkets (and their warehouses). Moreover, differently from independent stores or online-only stores, the biggest mass retail channel companies already have a capillary diffusion throughout the country, with clear advantages in terms of storage and logistics. On the other side, the intermediary role taken up by platforms permits them to simply 'match' producers and/or retailers with consumers without taking responsibility for storing food products. For instance, the platform Glovo is active in more than 20 countries worldwide and offers delivery for several food and nonfood products, among which groceries from supermarkets. Glovo must take care of only the logistics, but does not need a warehouse—though things have changed during the Covid-19 crisis (see below). Alternatively, Food Genuine, as seen above, connects consumers with local Italian producers (identified as having fewer than 50 employees), and lets them take care of the shipment.

**Box 1 Sketching Country Profiles**

An interesting result of our research concerns the almost unique configuration that online food provision appears to have in each of the examined countries. Overall, winner-take-all situations that have occurred in some sectors of the platform economy (e.g., accommodation) are not a feature of the European (online) food market, because multiple competitors clearly coexist within and across countries. Germany has the highest percentage of individuals purchasing food online (24% in 2019), and its online food market is characterised by a high number of independent stores selling a wide variety of products, perhaps reflecting the high number of small and medium enterprises (SMEs) that characterise its economic fabric (Eurostat, 2021, 2022). Although the few existing platforms (3) do in fact connect local producers and consumers, independent stores (50, 58.1%) are the ones that are mostly concerned with local food and that more generally report a higher number of commitments—especially concerning the environment and local food. Possibly, independent

(*continued*)

**Box 1 (continued)**

stores that also decide to sell online are managed by younger people more attentive to food sustainability and more accustomed to e-commerce infrastructures.

In Ireland, 9% of individuals purchased food online in 2019. Although this number has dramatically changed during and after the outbreak of Covid-19, the composition of its online food market is particularly interesting. Like Germany, Ireland presents a relatively high number of independent stores (39, 43.3%). Nevertheless, these shops do not seem interested in local food or more generally sustainability, as suggested by the low number of commitments in general. The food 'sustainability mission', so to say, is likely undertaken by the AFN, despite having a low range of food categories sold: the presence of 20 online AFN (22.2%) is the highest in absolute terms, a result which is particularly remarkable if we consider that Ireland has the smallest population among the countries considered. In fact, along with the only two platforms, AFN report the highest number of commitments in general and a particular attention towards local foods—though many of the online mass retailers also report it. This finding is particularly interesting, because it differs from the representation of AFN as being somewhat resistant to digitalisation, as was reported in other countries (e.g., Grasseni, 2013).

The Italian case is a rather puzzling one. Italy had the lowest number of online food shoppers in 2019 (4%), but an extremely lively market, with over 211 online food providers (3.6 per million people) and the highest number of platforms (22, 13.3%). The country presents also the highest number of online-only and independent stores—mirroring, as in the case of Germany, the diffusion of SMEs, particularly micro-enterprises (0–9 employees). As in the case of Ireland, AFN and platforms are the types of providers most concerned with sustainability and local food, although the latter is also a defining feature of the online-only shops, which are often entirely dedicated to the sale of typical gastronomic items—reflecting wider signals of gastronationalism and gastroregionalism that are typical of Italian food culture (DeSoucey, 2010; Lelieveldt, 2017; Oncini, 2019, 2020).

(*continued*)

**Box 1 (continued)**

Norway presents the highest number of online food providers once we take into account the size of the population (15 per million people). In 2019, 14% of its population purchased food online. As in Italy, Norway has a majority of online-only providers, a high number of platforms (13, 13.6%) and very few AFN. Most of the platforms connect local consumers and producers, and all of them report being committed to local food. Nevertheless, while the mass retail channel reports the highest number of commitments (well-being, environment, equity, local food, and nutrient needs) in general, other providers also notably maintain a high frequency of these commitments.

The Swedish profile is comparable to the Norwegian one, possibly indicating that Scandinavian countries are developing a similar online food market. Also, in this case, online-only shops are the majority (64.3%), and there are no AFN. Sweden presents the highest percentage of platforms dedicated to connecting producers and consumers locally. As in Norway however, it is the mass retail channel that reports more commitments, though they are less evenly distributed and most often refer solely to being committed to the environment more generally.

## 5 Online Grocery Shopping in Covid-19 Times: The Rise of On-Demand Grocery Delivery

Although the data in our possession were gathered two years before the appearance of Covid-19, it is plausible that the results shown above are still indicative of the main differences between the models we have identified. In particular, the fact that the online mass retail and the platform models are the only ones that can potentially guarantee a heterogeneous offer of food suggests that they could be the most effective substitutes for in-person grocery shopping. The outbreak of Covid-19 in early 2020 greatly affected food consumption and provision practices throughout Europe (Boons et al., 2020; Hirth et al., 2022; Oncini, 2021, 2022, 2023). Online food providers experienced tremendous growth in food sales and new customers, especially when the strictest lockdown measures were implemented. The diffusion of online grocery shopping in the first months after the virus outbreak remained however limited and did not cause a sudden shift in the food shopping practices of many families.

Nonetheless, the window of opportunity opened by the crisis created space for experimentation for both food retailers and consumers (Dannenberg et al., 2020), and throughout 2020 and 2021 tens of new on-demand grocery delivery services started operating all over Europe. These companies all depart from the models we have illustrated above and share the use of dark stores to stock the goods that need to be distributed. Dark stores are retail distribution centres or warehouses that are specifically designed to cater to online shoppers by fulfilling online orders for the local market. They serve online customers exclusively, and their purpose is to expedite the delivery of products to the customers' doorstep. Figure 7 below

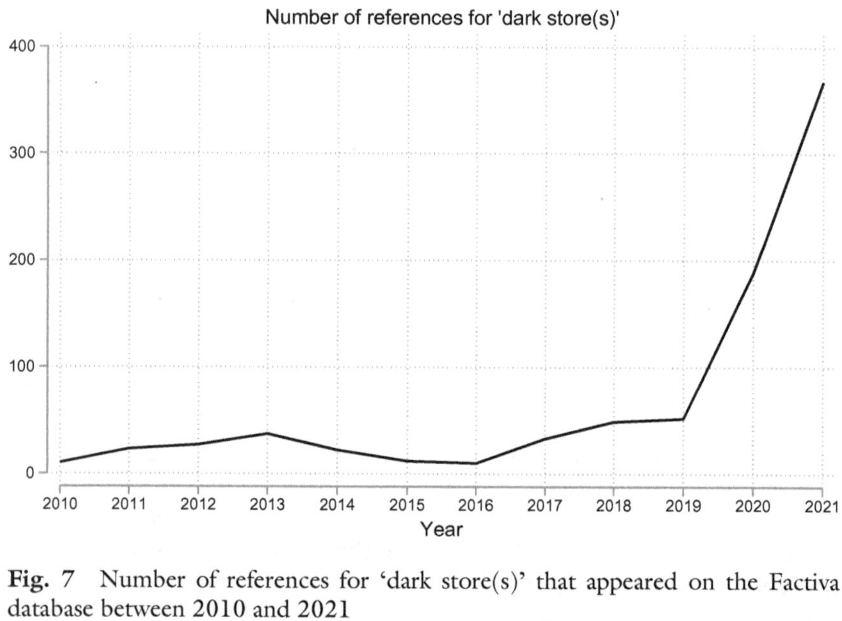


**Fig. 7** Number of references for 'dark store(s)' that appeared on the Factiva database between 2010 and 2021

used the Factiva[4] database to illustrate the number of times the words 'dark store' or 'dark stores' appeared in newspapers and journals over the last decade. Whereas until 2019 the terms appeared less than 50 times, in the following years, they clearly became more widely discussed, reaching 189 citations in 2019 and 368 in 2021. Companies such as Gorillas, Flink, Getir, Gopuff or Weezy are all based on dark stores and promise the instant delivery of food products (Butler, 2021). They can because dark stores are strategically placed throughout the city to guarantee delivery in 10 to 15 minutes through a fleet of riders, and in addition can offer shoppers a wide variety of food categories, including fresh products.

The emergence of these new actors is interesting because they seem to exploit the advantages of both pipelines and platforms simultaneously. On the one side, they could be considered as online-only pipelines, because they do not have an intermediary role between producers and consumers or between retailers and consumers. They do not offer a digital room for interaction as most famous platforms do. On the other side, the use of a fleet of riders carrying groceries from the closest dark store to the customer exploits the platform's capacity to offer a proximity match that permits decreasing the delivery time. In this sense, these companies exploit (i) the advantage of having as wide a range of food—especially popular convenience goods—as the mass retail channel without needing to manage the in-person shopping experience; and (ii) the advantage of relying on an extremely flexible and precarious workforce of riders that make possible the match between the closest dark store and the shopper.

It is too early to say whether these companies will become a fully-fledged alternative to the mass retail channel, whether the latter will be able to increasingly incorporate dark stores in the existing online food delivery practices, whether online food shopping will continue to be a residual shopping practice, or whether new options will appear. As I am writing, hybrid models seem to be on the rise—partly because of substantial investments from venture capitalists—and have made alliances with large corporate actors: Gorillas has entered into a partnership with Tesco (one of the largest UK mass retailers); other mass retailers are opening their own dark stores (Wallop, 2021); Amazon has opened new 'just walk in' stores where people shop directly from the shelves without checking

[4] The Factiva database aggregates information from more than 30,000 sources (such as newspapers, journals, etc.), and it is a widely used tool for research into journalistic content.

out (Nott, 2021). However, some commentators are arguing that the economic outlook is not as promising as these companies initially envisioned, as demonstrated by staff cuts and the closure of their services in some countries (e.g., Gorillas in Italy) (Mull, 2022).

## 6 Discussion and Conclusions

Building on the distinction between pipelines and platforms, this chapter outlined some findings that emerged from analysing a cross-country data set on the online food providers active in Germany, Ireland, Italy, Norway and Sweden in 2019, and advanced some reflections in light of the new on-demand grocery services that emerged during the pandemic. Although the Covid-19 pandemic could change, in the longer term, the face of food shopping as we know it, the classification presented above could be still used to distinguish online grocery shopping models in other countries. Clearly, the crisis affected the sector and created new opportunities for other business models to emerge, but many companies we identified as being in business three years ago are still running. Moreover, the distinction between pipelines and platforms continues to be useful, because it permits us to identify those (few) models that in each country aim to act as sole intermediaries. Despite marked differences between countries and between models, the data above indicate that platforms tend, on average, to present themselves as being more sustainable (similarly to AFN) but also as capable of offering a wider variety of food products (similar to the mass retail chains). The capacity of (some) platforms to reconnect producers and consumers while offering a wide variety of food products could seem appealing, especially in the face of an unprecedented sustainability crisis. Nevertheless, two years after the data collection, everything suggests that a food platform revolution is still far from taking place, at least not in the guise of the one that transformed the accommodation, the ICT and the mobility sectors. This is not to say that some food platforms did not increase their presence in the market: the Italian platform Cortilia, which aims to connect consumers and producers via a selected range of products having a short supply chain, has consistently increased its presence during the pandemic; similarly, the Food Assembly (see above) is in more countries now than it was two years ago. At present, however, these alternatives still seem to be attractive for a niche of customers or for a detour from routine practices, and not as substitutes for in-store grocery shopping.

During the pandemic, however, the increasing attraction of online grocery shopping created conditions for the growth of a new model that seems capable to exploit the efficiencies of *both* pipelines and platforms. On-demand grocery companies use, in addition, brick-and-mortar facilities closed to the public (dark stores) and a fleet of riders that make the match between shoppers and their closest warehouse possible. It is not by chance that *Glovo*, which initially operated as a platform matching shoppers with supermarkets, is now opening its own dark stores, and similarly, big players of the mass retail channel are now setting up dark stores and their own fleets of riders to compete with on-demand grocery companies, and/or entering into partnerships with them. Should this bifront Janus succeed, more than a food platform revolution we should be speaking of a food platform 'accommodation'. Against the optimistic vision of a new, potentially more sustainable, model of grocery shopping that could connect producers and consumers, thanks to the innovative potential of a digital intermediary, the rise of dark stores suggests that online grocery shopping will be similar to the traditional in-person model, but with added convenience for consumers.

The fast-paced evolution of the sector and the potential penetration of online grocery shopping as the new normal suggest lines of enquiry for theoretically informed research on both the provision and the purchasing sides. For one, it would be useful to reread and reorganise the multifaceted literature on online food provision and consumption following Dulsrud and Bygstad's (2022) identification of four research streams—the political, the economic, the technological and the sociocultural—that characterise research on digital ecosystems. Concurrently, several theoretical strands could help us better capture the coevolving nature of technology and society (see Chen's chapter in this collection for a thorough discussion).

Actor network theory (ANT) can improve our understanding of online grocery shopping by examining the complex web of relationships between technology, humans and nonhuman actors such as algorithms, data and delivery systems that shape the practice. Sustainability transition frameworks—and particularly the multi-level perspective (Geels, 2002; Geels & Schot, 2007)—could better illuminate how the sociotechnical system of food provision (i.e., the complex network of individual and collective actors, institutions, material artefacts and knowledge) is moving (or not) to the more sustainable modes of grocery shopping (Dannenberg et al.,

2020). Second, practice theory approaches could shed light on the reasons certain online food provisioning practices—and their engagements, materials and procedures—become (or not) engrained in the everyday routines of households and enable sustainable food provisioning (Dal Gobbo et al., 2022; Heidenstrøm & Hebrok, 2022; Samsioe & Fuentes, 2022; Southerton & Fuentes, 2022; Warde, 2005; see also Fuentes et al.'s chapter in this collection). In particular, practice theory could be particularly useful to understand the recent success of on-demand grocery shopping vis-à-vis the other models of food provision, and potentially its difficulties in moving from an emerging shopping practice to an established one (Fuentes et al., 2021).

A final note concerns the possibility of updating the 2019 PLATEFORM data set with new information given the recent changes in the industry. A follow-up study could help estimate the mortality rate of companies within each type of model, characterise on-demand grocery shopping companies, and, most importantly, identify whether the new model based on dark stores and fleets of riders has become more popular or if alternative models have emerged.

## References

Boons, F., Browne, A. L., Burgess, M., Ehgartner, U., Hirth, S., Hodson, M., Holmes, H., Hoolohan, C., MacGregor, S., Mcmeekin, A., Mylan, J., Oncini, F., Paterson, M., Rödl, M. B., Sharmina, M., Warde, A., Welch, D., Wieser, H., Yates, L., & Chunhua, Y. (2020). *Covid-19, changing social practices and the transition to sustainable consumption and production*. https://doi.org/10.13140/RG.2.2.18506.59840

Butler, S. (2021). Fast food: The new wave of delivery services bringing groceries in minutes. *The Guardian*. Retrieved February 20, 2023, from https://tinyurl.com/2p87m5y2

Dal Gobbo, A. D., Forno, F., & Magnani, N. (2022). Making "good food" more practicable? The reconfiguration of alternative food provisioning in the online world. *Sustainable Production and Consumption, 29*, 862–871. https://doi.org/10.1016/j.spc.2021.07.023

Dannenberg, P., Fuchs, M., Riedler, T., & Wiedemann, C. (2020). Digital transition by COVID-19 pandemic? The German food online retail. *Tijdschrift Voor Economische En Sociale Geografie, 111*(3), 543–560. https://doi.org/10.1111/tesg.12453

DeSoucey, M. (2010). Gastronationalism: Food traditions and authenticity politics in the European Union. *American Sociological Review, 75*(3), 432–455. https://doi.org/10.1177/0003122410372226

Dulsrud, A., & Bygstad, B. (2022). Digital ecosystems as social institutions: Exploring the role of consumption through four research streams of digital ecosystems. *Consumption and Society, 1*(1), 99–119. https://doi.org/10.1332/QUPH6141

Eurostat. (2021). *E-commerce statistics for individuals. Digital economy and society database*. Retrieved February 20, 2023, from https://tinyurl.com/4rpumm66

Eurostat. (2022). *Small and medium-sized enterprises (SMEs)*. Retrieved February 20, 2023, from https://tinyurl.com/3bze8fce

Forde, C., Stuart, M., Joyce, S., Oliver, L., Valizade, D., Alberti, G., Hardy, K., Trappmann, V., Umney, C., & Carson, C. (2017). *The social protection of workers in the platform economy* (IP/A/EMPL/2016-11). European Parliament. Retrieved February 20, 2023, from https://tinyurl.com/yru8c4kw

Forno, F., & Graziano, P. R. (2014). Sustainable community movement organisations. *Journal of Consumer Culture, 14*(2), 139–157. https://doi.org/10.1177/1469540514526225

Fuentes, C., Cegrell, O., & Vesterinen, J. (2021). Digitally enabling sustainable food shopping: App glitches, practice conflicts, and digital failure. *Journal of Retailing and Consumer Services, 61*, 102546. https://doi.org/10.1016/j.jretconser.2021.102546

Geels, F. W. (2002). Technological transitions as evolutionary reconfiguration processes: A multi-level perspective and a case-study. *Research Policy, 31*(8–9), 1257–1274. https://doi.org/10.1016/S0048-7333(02)00062-8

Geels, F. W., & Schot, J. (2007). Typology of sociotechnical transition pathways. *Research Policy, 36*(3), 399–417. https://doi.org/10.1016/j.respol.2007.01.003

Grasseni, C. (2013). *Beyond alternative food networks: Italy's solidarity purchase groups* (1st ed.). Bloomsbury Publishing.

Heidenstrøm, N., & Hebrok, M. (2022). Towards realizing the sustainability potential within digital food provisioning platforms: The case of meal box schemes and online grocery shopping in Norway. *Sustainable Production and Consumption, 29*, 831–850. https://doi.org/10.1016/j.spc.2021.06.030

Hirth, S., Oncini, F., Boons, F., & Doherty, B. (2022). Building back normal? An investigation of practice changes in the charitable and on-the-go food provision sectors through COVID-19. *Sustainability: Science, Practice and Policy, 18*(1), 410–427. https://doi.org/10.1080/15487733.2022.2076352

Lelieveldt, H. (2017). Comment on DeSoucey: Gastronationalism or gastroregionalism? *SSRN Electronic Journal*. https://doi.org/10.2139/ssrn.2984754

Mull, A. (2022). Of course instant groceries don't work. *The Atlantic*. Retrieved February 20, 2023, from https://tinyurl.com/ycyam5c

Nemes, G., Chiffoleau, Y., Zollet, S., Collison, M., Benedek, Z., Colantuono, F., Dulsrud, A., Fiore, M., Holtkamp, C., Kim, T.-Y., Korzun, M., Mesa-Manzano, R., Reckinger, R., Ruiz-Martínez, I., Smith, K., Tamura, N., Viteri, M. L., & Orbán, É. (2021). The impact of COVID-19 on alternative and local food systems and the potential for the sustainability transition: Insights from 13 countries. *Sustainable Production and Consumption, 28*, 591–599. https://doi.org/10.1016/j.spc.2021.06.022

Nott, G. (2021). Amazon opens third "Just Walk Out" store in London. *The Grocer*. Retrieved February 20, 2023, from https://tinyurl.com/yxfamt2d

Oncini, F. (2019). Feeding distinction: Economic and cultural capital in the making of food boundaries. *Poetics, 73*, 17–31. https://doi.org/10.1016/j.poetic.2019.02.002

Oncini, F. (2020). Cuisine, health and table manners: Food boundaries and forms of distinction among primary school children. *Sociology, 54*(3), 626–642. https://doi.org/10.1177/0038038519880087

Oncini, F. (2021). Food support provision in COVID-19 times: A mixed method study based in Greater Manchester. *Agriculture and Human Values, 38*(4), 1201–1213. https://doi.org/10.1007/s10460-021-10212-2

Oncini, F. (2022). Hunger bonds: Boundaries and bridges in the charitable food provision field. *Sociology*, online first. https://doi.org/10.1177/00380385221139024

Oncini, F. (2023). Charitable food provision as a strategic action field: Introducing a meso-level perspective on food support organizing. *Agriculture and Human Values*, online first. https://doi.org/10.1007/s10460-023-10480-0

Oncini, F., Bozzini, E., Forno, F., & Magnani, N. (2020). Towards food platforms? An analysis of online food provisioning services in Italy. *Geoforum, 114*, 172–180. https://doi.org/10.1016/j.geoforum.2020.06.004

Parker, G., Van Alstyne, M., & Choudary, S. P. (2016). *Platform revolution: How networked markets are transforming the economy and how to make them work for you* (1st ed.). W. W. Norton & Company.

Piroth, P., Ritter, M. S., & Rueger-Muck, E. (2020). Online grocery shopping adoption: Do personality traits matter? *British Food Journal, 122*(3), 957–975. https://doi.org/10.1108/BFJ-08-2019-0631

Samsioe, E., & Fuentes, C. (2022). Digitalizing shopping routines: Re-organizing household practices to enable sustainable food provisioning. *Sustainable Production and Consumption, 29*, 807–819. https://doi.org/10.1016/j.spc.2021.07.019

Southerton, D., & Fuentes, C. (2022). Special section: Digital platforms and sustainable food consumption transitions. *Sustainable Production and Consumption, 29*, 805–806. https://doi.org/10.1016/j.spc.2021.08.023

Vallas, S., & Schor, J. B. (2020). What do platforms do? Understanding the gig economy. *Annual Review of Sociology, 46*(1), 273–294. https://doi.org/10.1146/annurev-soc-121919-054857

van Dijck, J. (2021). Governing trust in European platform societies: Introduction to the special issue. *European Journal of Communication, 36*(4), 323–333. https://doi.org/10.1177/02673231211028378

Wallop, H. (2021). "We are democratising the right to laziness": The rise of on-demand grocery deliveries. *The Guardian*. https://tinyurl.com/2mz7ahdu

Warde, A. (2005). Consumption and theories of practice. *Journal of Consumer Culture, 5*(2), 131–153. https://doi.org/10.1177/1469540505053090

Yates, L. (2020). The politics of the platform economy: Understanding platform legitimation tactics. In M. Hodson et al. (Eds.), *Urban platforms and the future city*. Routledge.

Zuboff, S. (2019). *The age of surveillance capitalism: The fight for a human future at the new frontier of power* (1st ed.). Profile Books.

# Sustainable and Purchasing Behaviour of Online Food Shoppers: Survey Results from Italy, Ireland, and Germany

*Frode Alfnes, Silvia Biasetton, Margherita Brunori, and Mary McCarthy*

## 1 Introduction

Digitalisation is taking place in all dimensions of the food environment (Granheim et al., 2022; O'Neill et al., 2023; Ytreberg et al., 2023). The use of online platforms for food purchasing has significantly increased in recent years. The growth has been driven by early adopters using these platforms for some time and newcomers introduced to the platforms

F. Alfnes (✉)
Consumption Research Norway (SIFO), Oslo Metropolitan University, Oslo, Norway

School of Economics and Business, Norwegian University of Life Sciences, Oslo, Ås, Norway
e-mail: frode.alfnes@nmbu.no



A. Dulsrud, F. Forno (eds.), *Digital Food Provisioning in Times of Multiple Crises*, Consumption and Public Life,
https://doi.org/10.1007/978-3-031-46323-5_3

during the Covid-19 pandemic (Lu et al., 2022; Eger et al., 2021). This study presents the results of surveys conducted in January 2021 among alternative food network (AFN) users in Italy and Germany and online grocery store users in Ireland. The questionnaire covers online purchase frequencies of 14 product categories and self-assessments of 16 food sustainability indicators. We analyse the responses for four age and gender segments. Additionally, we examine the differences between early adopters and those who became regular users of online food platforms in 2020 by comparing demographics, purchasing behaviour, and purchase expectations for the future. The surveys were part of the ERA-NET project Plateforms and build on and complement the platform mapping (Oncini et al., 2020), method development (Heidenstrøm & Hebrok, 2021), practice conceptualisations (Samsioe & Fuentes, 2022), and ethnographically inspired in-depth interview studies (Heidenstrøm & Hebrok, 2021; Samsioe & Fuentes, 2022; O'Neill et al., 2022) conducted in the project. For a more in-depth study of the Irish survey respondents, see O'Neill et al. (2023).

Online food provisioning services, including online grocery stores, food delivery platforms, food waste reduction platforms, and AFNs, are now widely available in most developed countries (Oncini et al., 2020; Fuentes et al., 2021; Dal Gobbo et al., 2022). In addition to offering easy access to a diverse range of products, several of these services focus on sustainability in the food system, including concerns about the environmental impact of food production and consumption and the need to ensure food security and nutrition for present and future generations.

S. Biasetton
Department of Philosophy, Sociology, Pedagogy and Applied Psychology, University of Padua, Padua, Italy
e-mail: silvia.biasetton@phd.unipd.it

M. Brunori
Department of Law, University of Brescia, Brescia, Italy
e-mail: margherita.brunori@unibs.it

M. McCarthy
Department of Management and Marketing, University College Cork, Cork, Ireland
e-mail: M.McCarthy@ucc.ie

Online grocery stores and AFNs are very different forms of online food provisioning services. Online grocery stores typically offer a full assortment of products, are owned by leading grocery companies, and in many ways resemble traditional supermarkets. AFNs, on the other hand, is an umbrella term encompassing different groups and organisations, such as farmers' markets, community-supported agriculture, and solidarity purchasing groups. Scholars distinguished between different types of AFNs, such as consumer-driven or producer-driven ones, based on the actors' level of involvement and interest (Forssell & Lankoski, 2015; Pascucci et al., 2016; Preiss et al., 2017).

Studies of consumer food values typically find that the average consumer thinks food safety, nutrition, taste, and price are much more important than are the more altruistic values related to factors such as the environment, fairness, animal welfare, and origin (e.g., Lusk & Briggeman, 2009; Filimonau et al., 2022; Aprile & Punzo, 2022). When consumers are asked about sustainability, most studies find that consumers lack key knowledge on food-related sustainability topics (Austgulen et al., 2018; van Bussel et al., 2022). Studies focusing on sustainable-orientated food consumers, also referred to as green consumers, environmentally friendly consumers, or climate-friendly consumers, find that they tend to be female, whereas results with respect to other demographics are mixed (Verain et al., 2012).

Studies focusing on AFN consumers find that the use of these platforms is regularly associated with biospheric and altruistic values (Stern & Dietz, 1994; Chen, 2020). They concern themselves with environmental issues, preservation of biodiversity, food security, animal welfare, supporting local farmers and economies, sustainable production, and health and well-being (Corsi & Novelli, 2018; De Bernardi et al., 2019; Forssell & Lankoski, 2015; Turner & Hope, 2014). The short food chains found in most AFNs are associated with fresh food, minimising pollution caused by shipment, reducing costs by cutting intermediaries in the supply chain, preserving traditional food culture, and increasing food security (Forssell & Lankoski, 2015), associations that align well with the AFN consumers' value orientation.

Theories of Practice and Sustainable Consumption, proposed by Welch and Warde (2015), help us understand sustainable food practices. These theories suggest that sustainable consumption is influenced not only by individual choices but also by social conventions and structures. They argue that consumption practices are deeply embedded in daily routines, relationships, and broader societal norms and values, as studied in Plateform articles by Heidenstrøm and Hebrok (2021) and Samsioe and Fuentes (2022). The theories can be used to conceptualise, analyse, and make sense of empirical material and thereby to develop explanations for why people make certain food choices and for how social networks, cultural practices, and institutional structures influence sustainable food practices. They can also be used in studies of how community-based organisations play a role in promoting sustainable food options through education and raising awareness about them. Additionally, Shove (2003) argues that practices and their changes are linked to technological innovations. Online platforms, as networks that connect consumers and producers, are relevant to understanding how food purchasing practices change and how they affect people's attitudes and practices of sustainability (Heidenstrøm & Hebrok, 2021; Samsioe & Fuentes, 2022). Theories of practices inspired parts of our surveys, however, it should be noted that the survey format does not allow us to explore the complexity of consumption practices usually found in qualitative studies.

Our aim in this chapter is to profile online food shoppers across three countries, explore differences in their food purchasing and consumption behaviours, and identify trends evident across the three countries. We will focus on differences between groups within the three-country samples and general trends across the three samples. More specifically, we will examine the following research questions: How do online food purchase behaviours and practices vary across food categories in different types of online platforms? To what extent do age and gender differences influence online food purchase behaviour? How do sustainable food practice indicators vary by age and gender? To what extent did the Covid-19 pandemic impact the behaviours of the customer base for online food platforms? What are the expectations of online food consumers regarding their future online food shopping behaviour?

## 2 Surveys on Sustainable Practices of Online Platform Users

The ERA-NET project Plateforms developed an online questionnaire to gather data on individuals' experiences with online food platforms. It consisted of five sections: experiences with online food platforms, experiences with purchasing groceries online during the Covid-19 pandemic, desired characteristics of an ideal online platform for food acquisition, food consumption habits, and demographics. It took approximately ten minutes to complete the questionnaires. The data for this study were collected through surveys conducted in Italy, Germany, and Ireland in January 2021. The surveys were distributed through partnerships with online food platforms in Italy and Germany and an online survey provider in Ireland. This study will analyse responses from all sections except the ideal online platform section.

### *Platforms Used for the Surveys in the Three Countries*

The Irish survey includes respondents who were users of popular online supermarkets such as Tesco, Aldi, Dunnes, Lidl, and SuperValu. The survey was administered through a professional survey company. Conversely, the surveys for Italy and Germany were distributed through newsletters from AFNs and targeted users of vegetarian box delivery services, food assembly platforms, and consumer-driven platforms. Notably, the Irish sample revealed a relatively low percentage of users who reported using AFNs, at only 4%, indicating a low overlap between online supermarkets and AFNs.

The Italian survey includes 585 respondents from four platforms: Bioexpress, BuonMercato, L'Alveare che dice Sì, and FuoriMercato. Bioexpress is an online co-op that delivers boxes of organic fruits and vegetables from various regions of northern Italy. L'Alveare che dice Sì operates an online marketplace where local farmers and artisans can sell their products directly to consumers. BuonMercato and FuoriMercato are solidarity associations that offer members the option to book food through an online platform.

The German survey includes 148 respondents from the two platforms, Marktschwärmer and Märkische Kiste. Marktschwärmer is part of the same franchise chain as L'Alveare che dice Sì in Italy. Märkische Kiste is similar to Bioexpress in Italy as it is an online group that delivers boxes of organic fruits and vegetables. They also offer the option to purchase special boxes such as cheese, breakfast, and smoothie boxes.

### *Demographic Questions and Respondents' Share of Food Purchases Online*

We asked demographic questions to gather information about gender, age, and education. The age question includes four categories: under 30,

30–44, 45–59, and over 60. For education, we asked respondents to select the option that best describes their level of education from the following list: no formal education, primary education, secondary education, more than secondary education, and a university-level degree (such as a bachelor's degree, master's degree, or PhD). We also asked participants to report the share of their food grocery shopping that they did online before and after the start of the Covid-19 pandemic, using an 11-point scale where 0 represented nothing and 10 represented everything.

Table 1 presents the demographics of the three country samples. The samples are not representative of the general population. However, we will mainly be examining within-sample differences in this study.

The surveys employed different sampling strategies across the countries under study. In Ireland, the sample was recruited from a survey panel, and gender balance was used as a sampling criterion. Meanwhile, the German and Italian samples were acquired through convenience sampling from AFN platforms, and the majority of respondents in these samples were female. This can be ascribed to the conventional societal norm that grocery shopping is often a task assigned to women, further bolstered by the

**Table 1** Demographics and online purchase share of the three samples

| *Demographics* | *Italy* | *Germany* | *Ireland* |
|---|---|---|---|
| Women | 0.77 | 0.86 | 0.58 |
| Age groups | | | |
| < 30 | 0.02 | 0.03 | 0.20 |
| 30–44 | 0.27 | 0.39 | 0.42 |
| 45–59 | 0.46 | 0.44 | 0.24 |
| > 60 | 0.25 | 0.14 | 0.14 |
| Education | | | |
| No formal education | < 0.01 | 0.00 | < 0.01 |
| Primary education | 0.00 | 0.02 | 0.02 |
| Secondary education | 0.02 | 0.07 | 0.20 |
| More than secondary education | 0.32 | 0.19 | 0.22 |
| University-level degree | 0.66 | 0.72 | 0.56 |
| Share of food purchases online | | | |
| Before Covid-19 pandemic | 0.46 | 0.34 | 0.28 |
| Since the start of the pandemic | 0.68 | 0.49 | 0.55 |
| Size subsample | 585 | 148 | 335 |

Note: The variables Women, Age groups, and Education are dummy variables. Share of food purchases online is measured on an 11-point scale where 0 represents nothing and 10 represents everything. The variable is rescaled to shares between 0 and 1.

holistic food values championed by AFNs, which tend to resonate more strongly with female consumers.

### *Segmentation Question*

The participants are segmented on their nationality in all tables. We also divide the sample into four groups based on age and gender, using 45 years as the cut-off point. Additionally, we segment the sample based on when participants start using online platforms for food purchasing, with response options of 'Before 2010', '2010–2014', '2015–2019', '2020', and 'Do not use on a regular basis'. We create three sample segments: Early Adopters (EA) for those who start before the pandemic, 2020 Newcomers for those who began in 2020, and Not Regular Users (NRU) for those who do not use online platforms to purchase food regularly.

### *Purchasing Behaviour Questions*

We asked the participants about the frequency of their online purchases for 14 food categories, covering the most commonly consumed food. We asked, 'How often do you buy the following food items on online platforms for food acquisition?' The response options were: 1 = every week, 2 = 2–3 times a month, 3 = 2–3 times per three months, 4 = once or less per three months, and 5 = never. The 14 food categories in question are dairy products and eggs, meat and meat products, fish and seafood, bread and bakery products, fresh fruit and vegetables, oils, canned/bottled food products, seeds, dried fruits and pulses, nonalcoholic beverages, alcoholic beverages, coffee and tea, snacks, chocolate and candy, pasta, rice, and grains, and products for special dietary needs.

### *Food Sustainability Practice Questions*

The food sustainability section of the questionnaire consists of 32 questions about behaviours and practices that consumers associate with sustainable food behaviours and practices. Research in the Plateforms projects identified these behaviours and practices through ethnographically inspired in-depth interviews before we designed the questionnaire (Heidenstrøm & Hebrok, 2021; Samsioe & Fuentes, 2022; O'Neill et al., 2022). It is important to acknowledge that these questions reflect consumer-defined sustainable behaviours rather than providing objective sustainability measures. We will hereafter refer to these behaviours and practices as sustainable food practices, although some of them, for example, product and dietary choices, are not considered practices in the sense typically used in practice theory.

The 32 questions related to sustainable food practices are analysed and grouped in pairs, resulting in 16 food practice indicators. These indicators encompass a range of practices that consumers often associate with food sustainability, including cooking from scratch, avoiding red meat, avoiding food waste, eating or avoiding leftovers, eating a vegetarian diet, eating domestic products, eating after the best-before date, planned shopping, eating seasonally, eating fair trade products, using green transportation when shopping, keeping an overview of food at home, eating organic products, eating locally produced products, and prioritising quality over price.

The questions cover sustainable food practices in five sequences: organisation and planning (planning, having an overview), transportation (mode of transport used to purchase food), product choice attributes and purchases (purchasing seasonal, domestic, organic, high-quality, locally sourced, and fair trade products), dietary choices (consuming five servings of fruits and vegetables per day, avoiding meat, and following a vegetarian diet), and management of leftovers and waste (paying attention to best-before dates, using leftovers, and reducing food waste).

Table 2 presents the five sustainable food practice sequences, 16 sustainability food practice indicators, and the two items used to measure each indicator. The first item represents a positive sustainability practice, whereas the second represents a negative one. To calculate the score for each indicator, we summed the two items using reverse scoring for the second and then divided them by two. The resulting variables have a minimum score of 1 and a maximum score of 7, with higher scores indicating the more sustainable behaviours. It is important to note that these indicators may not measure unidimensional constructs, and some may have low correlation levels. For instance, planning in the sense of using a shopping list may have a low correlation with buying only what you intended.

### *Expectations for the Future Questions*

We asked participants about their previous, current, and expected online food shopping habits. The questions we used were 'What share of your food grocery shopping was done online before the Covid-19 pandemic?'

**Table 2** Food sustainability questions

| *Sequence* | *Indicators* | +/- | *Items* | *Correlation* | | |
|---|---|---|---|---|---|---|
| | | | | *IT* | *GE* | *IR* |
| Organisation and planning | Planning | + | I usually use a shopping list. | 0.14 | 0.25 | 0.17 |
| | | - | I frequently buy food products I had not planned to buy. | | | |
| | Overview | + | I have a good overview of the food I have at home. | 0.39 | 0.55 | 0.29 |
| | | - | I often find food in my kitchen that I had forgotten about. | | | |
| Transport | Transport | + | I usually walk, cycle, or take public transport for my food shopping. | 0.73 | 0.86 | 0.56 |
| | | - | I frequently drive to the place I do the food shopping. | | | |
| Product choice Attitudes and purchases | Seasonality | + | I mainly buy fresh fruits and vegetables that are in season. | 0.21 | 0.19 | 0.42 |
| | | - | I am not very interested in the seasonality of products. | | | |
| | Domestic | + | I prefer domestically produced food. | 0.38 | 0.29 | 0.34 |
| | | - | I rarely check the country of origin of the food I buy. | | | |
| | Organic | + | I prefer organic food. | 0.67 | 0.60 | 0.63 |
| | | - | I rarely buy organic food. | | | |
| | Quality | + | I value quality over price. | 0.44 | 0.34 | 0.39 |
| | | - | I usually choose the cheapest option. | | | |
| | Local | + | I prefer locally produced food. | 0.46 | 0.21 | 0.41 |
| | | - | I am not very interested in locally produced food. | | | |
| | Fairtrade | + | I prefer products where I can be sure that the producer has received a fair price. | 0.31 | 0.41 | 0.24 |
| | | - | I seldom choose Fairtrade products. | | | |

(*continued*)

**Table 2** (continued)

| *Sequence* | *Indicators* | +/- | *Items* | *Correlation* | | |
|---|---|---|---|---|---|---|
| | | | | *IT* | *GE* | *IR* |
| Dietary choice | Five a day | + | I eat at least five portions of fruits and vegetables a day. | 0.44 | 0.54 | 0.42 |
| | | - | I eat very little fruit and vegetables. | | | |
| | Avoid meat | + | I try to avoid consuming red meat. | 0.48 | 0.49 | 0.53 |
| | | - | I have meat-based meals most days. | | | |
| | Vegetarian | + | I eat vegetarian meals at least every second day. | 0.72 | 0.78 | 0.49 |
| | | - | I seldom eat vegetarian meals. | | | |
| | Cook from scratch | + | I cook most of my meals from scratch. | 0.37 | 0.60 | 0.38 |
| | | - | I frequently buy ready-made food. | | | |
| Leftovers and waste | Best before | + | I sometimes eat food even if it has past the best-before date. | 0.61 | 0.67 | 0.51 |
| | | - | I always throw away food that is past the best-before date. | | | |
| | Leftover | + | I store and use leftover food. | 0.52 | 0.59 | 0.52 |
| | | - | I rarely eat leftover food. | | | |
| | Food waste | + | I try to avoid creating food waste. | 0.34 | 0.29 | 0.22 |
| | | - | I sometimes find food in my kitchen that has gone off. | | | |

Note: Respondents were asked to rate how well the following statements described them as food consumers on a 7-point scale, where 1 is 'Does not describe me at all' and 7 is 'Describes me very well'. The variables representing the traits were created by reversing any negative items, summing the two scores, and dividing by two, resulting in variables with a minimum score of 1 and a maximum score of 7

'What share of your food grocery shopping was done online since the start of the Covid-19 pandemic?', and 'Looking a few years ahead, when the Covid-19 pandemic is over, do you think the share of your food grocery shopping done online will increase, remain unchanged, or decrease compared to now?' For the first two questions, we use an 11-point scale where 0 is nothing, and 10 is everything. For the last question, the response options are 'Increase', 'Remain unchanged', and 'Decrease'.

## 3 Results and Discussion

In this section, we present and discuss the results of our surveys of food platform users. We first provide an overview of the frequency of purchases for the various food categories across the three country samples. Second, we examine differences in the 16 sustainable food practice indicators among age and gender segments in the three country samples. Finally, we investigate differences between early adopters, the 2020 newcomers, and non-regular users regarding purchase frequency, sustainable behaviour, and expectations about online purchases after the pandemic.

### *Online Product Purchase Frequencies by Age and Gender Segments*

Table 3 presents the share of the four age-gender segments in the three country samples that purchase each of the 14 food categories more than once monthly on online platforms.

Table 3 shows that in the Italian and German samples, almost all respondents purchase fresh fruits and vegetables, and about half purchase dairy products and eggs from online platforms monthly. For the other categories, the share of monthly buyers is less than a third of the sample. This may be due to the focus of the AFN online platforms we used in the Italian and German recruitment and suggests that these platforms are the source of some but not all types of food.

In the Irish sample, the products are more evenly distributed, with eight categories having more than 40% of respondents buying at least monthly and no category having more than 55% of respondents buying monthly. Like in Italy and Germany, fresh fruits and vegetables have the highest monthly buyers at 53%. At the other end of the scale, we also note that 25% of the Irish sample indicate that they buy products for special dietary needs online at least monthly. We believe the differences in purchase frequency between the three countries at least partly result from the characteristics of the online platforms used by the three country samples.

When examining the disparities in food purchasing behaviour by age and gender, the results are not straightforward. In the Irish sample, individuals under the age of 45 constituted a majority of 62% of the respondents and emerged as the most frequent buyers across all food categories. In contrast, the corresponding figures for the Italian and German samples were 28% and 42%, respectively. In Italy, younger users of online platforms

**Table 3** Share that purchases product categories more than once monthly on online platforms

| *Country* | *Italy* | | | | *Germany* | | | | *Ireland* | | | |
|---|---|---|---|---|---|---|---|---|---|---|---|---|
| *Gender/Age* | *Female* | | *Male* | | *Female* | | *Male* | | *Female* | | *Male* | |
| *Food categories* | –45 | 45+ | –45 | 45+ | –45 | 45+ | –45 | 45+ | –45 | 45+ | –45 | 45+ |
| Dairy products and eggs | 0.51 | 0.47 | 0.51 | 0.44 | 0.53 | 0.55 | 0.64 | 0.60 | **0.59** | **0.49** | **0.97** | **0.87** |
| Meat and meat products | 0.26 | 0.24 | 0.38 | 0.28 | 0.24 | 0.36 | 0.36 | 0.40 | 0.56 | 0.46 | 0.51 | 0.44 |
| Fish and seafood | 0.22 | 0.16 | 0.24 | 0.23 | **0.04** | **0.08** | **0.18** | **0.30** | 0.61 | 0.47 | 0.30 | 0.28 |
| Bread and bakery products | 0.38 | 0.30 | 0.30 | 0.28 | **0.35** | **0.42** | **0.09** | **0.60** | **0.60** | **0.49** | **0.38** | **0.28** |
| Fresh fruit and vegetables | **0.95** | **0.93** | **0.97** | **0.87** | 0.82 | 0.91 | 0.91 | 1.00 | **0.50** | **0.33** | **0.35** | **0.33** |
| Oils | **0.13** | **0.06** | **0.16** | **0.07** | **0.04** | **0.07** | **0.00** | **0.30** | **0.41** | **0.27** | **0.24** | **0.28** |
| Canned / bottled food products | 0.23 | 0.24 | 0.24 | 0.28 | 0.24 | 0.17 | 0.27 | 0.50 | **0.51** | **0.40** | **0.14** | **0.20** |
| Seeds, dried fruits, and pulses | 0.28 | 0.21 | 0.27 | 0.22 | 0.10 | 0.08 | 0.09 | 0.10 | **0.43** | **0.36** | **0.24** | **0.23** |
| Nonalcoholic beverages | **0.20** | **0.14** | **0.08** | **0.09** | 0.14 | 0.17 | 0.18 | 0.40 | **0.30** | **0.23** | **0.27** | **0.22** |
| Alcoholic beverages | **0.16** | **0.06** | **0.14** | **0.05** | 0.06 | 0.05 | 0.00 | 0.20 | **0.50** | **0.30** | **0.08** | **0.09** |
| Coffee and tea | 0.19 | 0.14 | 0.16 | 0.13 | 0.10 | 0.07 | 0.00 | 0.20 | **0.49** | **0.33** | **0.16** | **0.13** |
| Snacks, chocolate, and candies | **0.23** | **0.12** | **0.14** | **0.20** | 0.10 | 0.11 | 0.09 | 0.10 | 0.40 | 0.23 | 0.14 | 0.05 |
| Pasta, rice, and grains | 0.36 | 0.27 | 0.35 | 0.33 | 0.10 | 0.11 | 0.09 | 0.20 | **0.33** | **0.17** | **0.16** | **0.07** |
| Products for special dietary needs | 0.11 | 0.07 | 0.11 | 0.09 | 0.04 | 0.07 | 0.09 | 0.00 | **0.30** | **0.13** | **0.11** | **0.09** |
| Size subsample | 129 | 324 | 37 | 95 | 51 | 76 | 11 | 10 | 138 | 57 | 70 | 70 |

Note: The four segments per country are females less than 45, females 45 or older, males less than 45, and males 45 or older. Bold numbers indicate significant differences between segments for a food category within a country. The original scale for the frequency of online purchases was: 1 = every week, 2 = 2–3 times a month, 3 = 2–3 times per three months, 4 = once or less per three months, and 5 = never. The dummy used to calculate the shares equals 1 for those responding 1 or 2 on the original scale emerged as the most frequent buyers in 13 out of 14 categories, whereas the data from Germany provided less conclusive evidence.

Online supermarkets in Ireland offer convenience and timesaving benefits, and the high proportion of young customers suggests that the under-45 age group most values these benefits. In contrast, AFNs are often viewed as less convenient and timesaving and tend to attract an older clientele. The online purchase frequencies in Ireland and Italy indicate that the under-45 age group has integrated online food shopping into their routines more than older online shoppers have, leading to higher purchase frequencies across all types of food.

### *Sustainability Food Practice Indicators by Age and Gender Segments*

Table 4 presents the results for the sustainable food practice indicators of the four age-gender segments in the three country samples. The 16 indicators are based on the questions presented in Table 2, with 1 indicating a low level and 7 a high level of sustainability behaviour.

As presented in Table 2, we divide the 16 sustainable food practices into five sequences: organisation and planning, transportation, product choice attributes and purchases, dietary choices, and management of leftovers and waste.

From Table 4, we first note that respondents recruited from AFN platforms in Italy and Germany displayed a stronger sustainable orientation than did Irish respondents recruited from online grocery store shoppers. The overall weighted average for the Italian and German respondents was 5.66 and 5.73, respectively, whereas the overall average for the Irish respondents was 4.60. This difference holds for all four age-gender segments and all five sustainability areas. For example, the respondents in Italy and Germany score themselves higher than the Irish respondents in managing leftovers and waste, with scores of 5.79, 5.94, and 4.92, respectively. This difference may be attributed to AFNs attracting customers more interested in sustainability than the average consumer is (see, e.g., Corsi & Novelli, 2018; De Bernardi et al., 2019), whereas there is likely not as strong a sustainability orientation among online grocery store consumers.

Second, in all three countries, women score higher than men on the sustainability indicators, whereas there are only tiny differences based on age. We find the most significant gender differences in dietary choices,

**Table 4** Sustainability food practice indicators by age and gender segments

| *Subsamples* | *Italy* | | | | *Germany* | | | | *Ireland* | | | |
|---|---|---|---|---|---|---|---|---|---|---|---|---|
| *Gender/ Age* | *Female* | | *Male* | | *Female* | | *Male* | | *Female* | | *Male* | |
| *Indicators* | *−45* | *45* | *−45* | *45* | *−45* | *45* | *−45* | *45* | *−45* | *45* | *−45* | *45* |
| Planning | 5.66 | 5.61 | 5.71 | 5.44 | 5.13 | 5.35 | 4.86 | 5.00 | 4.36 | 4.40 | 4.49 | 4.61 |
| St.d. | 0.10 | 0.07 | 0.16 | 0.12 | 0.18 | 0.14 | 0.44 | 0.44 | 0.11 | 0.19 | 0.13 | 0.16 |
| Overview | 6.03 | 6.00 | 5.92 | 5.89 | 5.78 | 6.01 | 5.86 | 5.70 | **5.01** | **5.59** | **4.93** | **5.34** |
| St.d. | 0.09 | 0.06 | 0.15 | 0.11 | 0.13 | 0.11 | 0.36 | 0.52 | 0.10 | 0.15 | 0.15 | 0.15 |
| Transport | **4.51** | **4.13** | **3.83** | **3.76** | 5.00 | 5.16 | 5.45 | 5.85 | 3.69 | 3.10 | 3.77 | 3.67 |
| St.d. | 0.20 | 0.12 | 0.37 | 0.20 | 0.30 | 0.26 | 0.74 | 0.37 | 0.17 | 0.28 | 0.22 | 0.22 |
| Seasonality | **6.54** | **6.46** | **6.33** | **6.17** | **5.99** | **6.10** | **5.55** | **6.30** | **4.91** | **5.50** | **4.76** | **5.11** |
| St.d. | 0.07 | 0.05 | 0.14 | 0.10 | 0.14 | 0.11 | 0.20 | 0.26 | 0.12 | 0.15 | 0.15 | 0.15 |
| Domestic | **5.88** | **6.09** | **5.53** | **6.02** | **5.87** | **6.15** | **5.95** | **5.95** | **4.40** | **5.25** | **4.59** | **4.93** |
| St.d. | 0.12 | 0.07 | 0.21 | 0.12 | 0.12 | 0.12 | 0.22 | 0.29 | 0.11 | 0.17 | 0.15 | 0.15 |
| Organic | 6.19 | 6.07 | 5.85 | 6.03 | **6.32** | **6.37** | **5.55** | **6.15** | 4.26 | 4.18 | 4.37 | 4.05 |
| St.d. | 0.09 | 0.07 | 0.21 | 0.13 | 0.10 | 0.11 | 0.37 | 0.39 | 0.14 | 0.23 | 0.19 | 0.20 |
| Quality | 5.76 | 5.80 | 5.60 | 5.84 | **6.05** | **6.29** | **5.59** | **6.20** | **4.53** | **4.87** | **4.29** | **4.67** |
| St.d. | 0.09 | 0.06 | 0.18 | 0.11 | 0.13 | 0.10 | 0.23 | 0.39 | 0.11 | 0.16 | 0.14 | 0.15 |
| Local | **6.22** | **6.07** | **5.76** | **5.93** | 6.25 | 6.27 | 5.86 | 6.10 | **4.84** | **5.44** | **4.86** | **5.11** |
| St.d. | 0.09 | 0.07 | 0.19 | 0.11 | 0.14 | 0.12 | 0.26 | 0.41 | 0.11 | 0.17 | 0.16 | 0.14 |
| Fairtrade | 5.58 | 5.52 | 5.18 | 5.52 | 5.87 | 5.95 | 5.50 | 6.10 | 4.50 | 4.78 | 4.41 | 4.64 |
| St.d. | 0.12 | 0.07 | 0.19 | 0.12 | 0.14 | 0.12 | 0.30 | 0.31 | 0.10 | 0.16 | 0.13 | 0.13 |
| Five a day | **5.82** | **5.96** | **5.47** | **5.32** | 5.32 | 5.40 | 4.86 | 5.05 | **4.74** | **5.07** | **4.37** | **5.03** |
| St.d. | 0.11 | 0.06 | 0.22 | 0.14 | 0.15 | 0.13 | 0.55 | 0.33 | 0.12 | 0.20 | 0.14 | 0.15 |
| Avoid meat | **5.52** | **5.59** | **4.99** | **5.33** | 5.14 | 5.30 | 4.41 | 4.80 | **4.14** | **3.79** | **3.68** | **3.43** |
| St.d. | 0.13 | 0.08 | 0.25 | 0.15 | 0.23 | 0.16 | 0.69 | 0.56 | 0.15 | 0.24 | 0.18 | 0.20 |
| Vegetarian | **5.68** | **5.21** | **5.15** | **4.63** | 6.03 | 5.91 | 5.32 | 5.40 | **4.09** | **3.80** | **3.36** | **3.26** |
| St.d. | 0.14 | 0.10 | 0.32 | 0.21 | 0.21 | 0.15 | 0.52 | 0.51 | 0.15 | 0.25 | 0.16 | 0.20 |
| From scratch | **6.14** | **6.16** | **5.81** | **5.76** | **5.86** | **6.34** | **5.09** | **5.95** | **5.02** | **5.69** | **4.69** | **5.17** |
| St.d. | 0.09 | 0.06 | 0.18 | 0.12 | 0.16 | 0.09 | 0.37 | 0.43 | 0.12 | 0.17 | 0.14 | 0.15 |
| Best before | **5.30** | **5.33** | **4.89** | **4.90** | 6.34 | 6.09 | 6.18 | 6.10 | 4.20 | 4.18 | 4.26 | 4.39 |
| St.d. | 0.15 | 0.10 | 0.26 | 0.19 | 0.15 | 0.13 | 0.26 | 0.59 | 0.14 | 0.24 | 0.18 | 0.19 |
| Leftover | **6.29** | **6.28** | **6.07** | **5.96** | 5.93 | 6.19 | 5.86 | 5.70 | **5.21** | **5.83** | **4.87** | **5.13** |
| St.d. | 0.09 | 0.06 | 0.23 | 0.13 | 0.21 | 0.10 | 0.35 | 0.60 | 0.12 | 0.17 | 0.17 | 0.17 |
| Food waste | 6.03 | 6.19 | 6.08 | 6.12 | 5.62 | 5.64 | 5.64 | 6.00 | **4.91** | **5.61** | **5.06** | **5.40** |
| St.d. | 0.08 | 0.05 | 0.15 | 0.09 | 0.14 | 0.13 | 0.34 | 0.47 | 0.11 | 0.14 | 0.14 | 0.14 |
| Size subsample | 128 | 321 | 37 | 93 | 51 | 76 | 11 | 10 | 138 | 57 | 70 | 70 |

Note: The 16 characteristics are based on the questions presented in Table 2, with a scale of 1–7, where 1 does not describe the respondents at all and 7 describes them very well. We used a Kruskal–Wallis equality-of-populations rank test to test for significant differences between the segments for each country and characteristic. Bold numbers indicate significant differences within one indicator in one country with overall scores of 5.32 for women and 4.85 for men. For all four dietary choice indicators, women score themselves higher than men do. The most significant gender difference is in avoiding meat and eating a vegetarian diet among respondents below 45. The gender difference aligns with previous research on meat consumption (e.g., Austgulen et al., 2018).

Third, we find the largest age differences in product choice attributes and purchases, with overall scores of 5.43 and 5.67 for consumers below and above 45, respectively. We find higher scores for the older respondents on all six indicators, with purchasing domestic (5.37 vs 5.73) and high quality (5.30 vs 5.61) having the largest differences.

### *Early Adopters, 2020 Newcomers, and Non-regular Users*

Table 5 presents the demographics of early adopters, 2020 newcomers, and non-regular users in the three country samples.

Based on the results in Table 5, approximately 20% of regular users of AFNs in Italy and Germany began using these networks in 2020. In contrast, more than half of regular users of online grocery stores in Ireland started to shop regularly online in 2020. The lower proportion of new users in the Italian and German samples may partly be attributed to the fact that these respondents were recruited through the AFNs themselves, and new users may be less likely to respond to a survey from the platform. In contrast, the Irish sample was recruited through a professional survey panel without a direct connection to the online food platform.

**Table 5** Demographics of early adopters, 2020 newcomers, and non-regular users

| *Demographics* | *Italy* | | | *Germany* | | | *Ireland* | | |
|---|---|---|---|---|---|---|---|---|---|
| *Segments* | *EA* | *2020* | *NRU* | *EA* | *2020* | *NRU* | *EA* | *2020* | *NRU* |
| Women | 0.78 | 0.75 | 0.84 | 0.86 | 0.81 | 1.00 | 0.57 | 0.56 | 0.67 |
| Age groups | | | | | | | | | |
| < 30 | 0.01 | 0.06 | 0.05 | 0.01 | 0.10 | 0.00 | 0.19 | 0.20 | 0.19 |
| 30–44 | 0.25 | 0.38 | 0.16 | 0.44 | 0.26 | 0.29 | 0.45 | 0.44 | 0.28 |
| 45–59 | 0.50 | 0.31 | 0.42 | 0.39 | 0.58 | 0.57 | 0.26 | 0.20 | 0.30 |
| > 60 | 0.25 | 0.26 | 0.37 | 0.16 | 0.06 | 0.14 | 0.09 | 0.16 | 0.22 |
| University-level degree | 0.65 | 0.69 | 0.63 | 0.75 | 0.61 | 0.71 | 0.54 | 0.61 | 0.46 |
| Size subsample | 457 | 109 | 19 | 110 | 31 | 7 | 138 | 151 | 46 |

Note: We here use a dummy variable for university-level degrees (BA, MA, PhD)

We conduct chi-squared tests to investigate the association between segments and demographics. Results show that there is no significant association between segments and the demographics of gender and education across all three countries. This lack of a significant association suggests that the newcomers who began shopping online during the pandemic do not differ with respect to gender and education from the pre-pandemic users in all three countries. However, our analysis reveals a significant association between segments and age in Italy and Germany. In Italy, we observe that most early adopters fall within the 45–59 age range, with only 25% being under 45 years old. This could mean that older individuals are more willing to try out the less convenient AFNs, possibly due to a greater interest in local and higher-quality food and more time for purchasing food. In contrast, the highest number of new Italian users during the pandemic is in the 30–44 age range, with 43% being less than 45. This difference suggests that the new users of the Italian alternative food platforms were younger than the existing users. In Germany, the results are different. The largest age segment among early adopters is 30–44, whereas the largest segment among new users is 45–59. Additionally, 55% of early adopters of the two platforms in Germany were above 45, whereas 64% of new users were above 45. Hence, there are indications of a small change in the age composition of users during the pandemic for both the Italian and German platforms.

Table 6 presents the shares of early adopters, 2020 newcomers, and non-regular users that purchase online from each product category more than once monthly.

We can see that the purchasing frequencies of the early adaptors and the 2020 newcomers are very similar in all three countries. Low purchase frequencies of the non-regular users mainly drive the significant difference found in the tests. For the Italian and German samples, we can see that around 90% of both early adopters and 2022 newcomers say they buy fresh fruits and vegetables online at least monthly. For dairy products and eggs, meat and meat products, and bread and bakery products approximately 50%, 30% and 30%, respectively, of respondents in both Italy and Germany say they purchase these categories online at least once monthly. We see different purchasing frequencies for the Irish sample than for the Italian and German samples, but not between early adopters and the 2020 newcomers. The differences we see between the countries can likely, to a large degree, be ascribed to the differences in the product selection in the online platforms they use. The Irish online grocery stores have a much

**Table 6** Share that purchases product categories more than once monthly on online platforms

| *Samples* | *Italy* | | | *Germany* | | | *Ireland* | | |
|---|---|---|---|---|---|---|---|---|---|
| *Segments* | *EA* | *2020* | *NRU* | *EA* | *2020* | *NRU* | *EA* | *2020* | *NRU* |
| Food categories | | | | | | | | | |
| Dairy products and eggs | **0.49** | **0.50** | **0.05** | 0.56 | 0.61 | 0.14 | **0.54** | **0.59** | **0.20** |
| Meat and meat products | 0.27 | 0.26 | 0.05 | 0.32 | 0.32 | 0.29 | **0.51** | **0.59** | **0.22** |
| Fish and seafood | **0.19** | **0.21** | **0.00** | 0.08 | 0.13 | 0.00 | **0.46** | **0.44** | **0.11** |
| Bread and bakery products | **0.31** | **0.37** | **0.11** | 0.36 | 0.52 | 0.14 | **0.52** | **0.58** | **0.22** |
| Fresh fruit and vegetables | **0.95** | **0.92** | **0.53** | **0.92** | **0.87** | **0.43** | **0.57** | **0.59** | **0.22** |
| Oils | 0.08 | 0.09 | 0.00 | 0.06 | 0.10 | 0.00 | 0.30 | 0.31 | 0.17 |
| Canned /bottled food products | **0.25** | **0.28** | **0.00** | 0.25 | 0.19 | 0.00 | **0.39** | **0.44** | **0.24** |
| Seeds, dried fruits, and pulses | 0.24 | 0.23 | 0.05 | 0.09 | 0.10 | 0.00 | **0.36** | **0.30** | **0.11** |
| Nonalcoholic beverages | 0.15 | 0.12 | 0.11 | 0.18 | 0.16 | 0.14 | **0.41** | **0.43** | **0.17** |
| Alcoholic beverages | 0.09 | 0.08 | 0.00 | 0.08 | 0.00 | 0.00 | **0.35** | **0.34** | **0.13** |
| Coffee and tea | 0.16 | 0.15 | 0.00 | **0.08** | **0.03** | **0.29** | **0.43** | **0.46** | **0.17** |
| Snacks, chocolate, and candies | **0.16** | **0.21** | **0.00** | 0.10 | 0.13 | 0.00 | **0.50** | **0.51** | **0.20** |
| Pasta, rice, and grains | **0.32** | **0.30** | **0.00** | 0.11 | 0.10 | 0.14 | **0.41** | **0.48** | **0.20** |
| Products for special dietary needs | 0.09 | 0.06 | 0.00 | 0.06 | 0.03 | 0.00 | **0.29** | **0.27** | **0.09** |
| Size subsample | 454 | 109 | 19 | 110 | 31 | 7 | 138 | 151 | 46 |

Note: The three segments per country are early adopters (EA), 2020 newcomers (2020), and non-regular users (NRU). Bold numbers indicate significant differences between segments for a food category within a country. The original scale for the frequency of online purchases was: 1 = every week, 2 = 2–3 times a month, 3 = 2–3 times per three months, 4 = once or less per three months, and 5 = never. The dummy used to calculate the shares equals 1 for those responding 1 or 2 on the original scale

wider product selection than the AFNs in Italy and Germany do. The latter focuses more on the basic products coming directly from the primary producers, like fresh fruits and vegetables.

Segment analysis of online food platform use over time is presented in Table 7. As expected, for all three countries, respondents who said they were regular users of online food platforms before 2020 reported a significantly higher share of online food purchases before the pandemic than the other two segments did. All the segments across the three countries report an increased share of food purchased online during the pandemic, with the newcomers having the largest increases. Those stating that they became

**Table 7** Online food platform use over time by segments

| *Demographics* | *Italy* | | | *Germany* | | | *Ireland* | | |
|---|---|---|---|---|---|---|---|---|---|
| *Segment* | *EA* | *2020* | *NRU* | *EA* | *2020* | *NRU* | *EA* | *2020* | *NRU* |
| Share online before[ad] | 0.53 | 0.22 | 0.30 | 0.40 | 0.13 | 0.26 | 0.45 | 0.18 | 0.12 |
| Share online in panademic[bd] | 0.69 | 0.69 | 0.48 | 0.49 | 0.49 | 0.40 | 0.57 | 0.60 | 0.29 |
| Expected change after[c] | | | | | | | | | |
| Increase | 0.27 | 0.22 | 0.32 | 0.25 | 0.45 | 0.29 | 0.38 | 0.34 | 0.22 |
| Remain unchanged | 0.63 | 0.62 | 0.47 | 0.69 | 0.62 | 0.71 | 0.51 | 0.34 | 0.57 |
| Decrease | 0.10 | 0.16 | 0.21 | 0.05 | 0.03 | 0.00 | 0.10 | 0.32 | 0.22 |
| Size subsample | 454 | 109 | 19 | 110 | 31 | 7 | 138 | 151 | 46 |

[a] What share of your food grocery shopping was done online before the Covid-19 pandemic?

[b] What share of your food grocery shopping was done online since the start of the Covid-19 pandemic?

[c] Look a few years ahead when the Covid-19 pandemic is over. Compared to now, do you think the share of your food grocery shopping done online will increase, remain unchanged, or decrease?

[d] Share of food purchases online is measured on an 11-point scale where 0 represents nothing and 10 represents everything. The variable is rescaled to shares between 0 and 1

regular online food shoppers in 2020 buy a similar share of their food online as those who have used online stores longer.

In the coming years, most online food consumers anticipate maintaining or even increasing their level of online food shopping. Only a small percentage of early adopters expect to decrease their online food shopping in the future. However, the expectations of 2020 newcomers and non-regular online shoppers are more varied. Whereas a third of Irish newcomers anticipate reducing online food purchases post-pandemic, a relatively smaller percentage of German and Italian newcomers share this expectation. The majority in all groups anticipate maintaining or further increasing their online food shopping in the future. This means that many of the new customers who started using online platforms to purchase food during the pandemic are likely to continue, and that online stores are more likely than AFNs to experience a drop in sales after the pandemic.

## 4 Concluding Remarks

In conclusion, this chapter presented the results of our surveys of online food platform users in Italy, Ireland, and Germany in January 2021. Our findings suggest that the Covid-19 lockdown has been positive for online food suppliers, with many new consumers using their services. Notably, most of these new customers plan to continue using the service, having experienced its many benefits. The purchasing frequencies among early adopters and newcomers to online food platforms in 2020 were similar across all three countries. The differences we observed in purchasing frequencies across the countries can largely be attributed to the differences in product selection available on the online platforms used for recruitment. The Italian and German samples recruited through AFNs showed the highest purchase frequencies for basic products such as fresh fruits and vegetables, whereas the Irish sample recruited from online grocery store users showed a high online purchase frequency for the wider range of products available in online stores.

Furthermore, we found that the respondents recruited from AFN platforms in Italy and Germany had a stronger inclination towards sustainable food behaviours and practices than did those recruited from online grocery store shoppers in Ireland. AFN platforms typically promote local, organic, and sustainable food production and consumption practices and thus may attract consumers who are already interested in sustainability. In contrast, online grocery stores in Ireland may primarily appeal to consumers seeking convenience and timesaving benefits rather than to those prioritising sustainability.

Our analysis of sustainable food behaviours and practices was inspired by the Plateform studies using theories of practice. However, our sustainable food practice indicators include a wider set of indicators than what is usually considered practices. Furthermore, surveys are usually not preferred to study the complexity of consumption practices. With that in mind, our sustainable food practice indicators were operationalised across five practice elements: organisation and planning, transportation, product choice attributes and purchases, dietary choices, and management of leftovers and waste. Our investigation of respondents from AFNs revealed higher scores across all five elements, which suggests that such platforms do not merely serve as a source of food but also promote sustainable food choices as part of a broader sustainable consumption practice. These findings align with existing scholarship on sustainable food practices, which posits that such practices involve multiple interrelated dimensions of everyday life. Our findings also suggest that sustainable food practices involve various interrelated actions, and AFN platforms may be effective in promoting these practices to their users.

Our research provides valuable insights into the current state of online food provisioning services and the purchasing behaviours of their users.

The data collected in this study can be used by policymakers and online platform providers to understand consumer behaviour and preferences and to develop strategies that promote sustainable food choices.

Our research has pointed towards several strategies that AFNs can employ to promote sustainable food practices and increase sales of sustainable food. Their customers buy only a few food categories online. AFNs can expand their online product offerings to include a wider range of sustainable food from different food categories, achieved through a wider collaboration with sustainable food suppliers offering high-quality and environmentally friendly food products. Finding ways to reach out to young consumers should be an important target for AFNs in the coming years. Convenience is a crucial factor for this segment, which AFNs have been less successful in attracting. Making it easier for the customers to get the products they want at the time they want them should be a priority. AFNs are uniquely positioned to provide information on sustainable food practices to sustainability-oriented consumers. By providing more information, AFNs have the potential to increase the number of sustainable food practices their customers follow. Another strategy that should be tested is green online nudges, which can be seen in many online grocery stores (Ytreberg et al., 2023). These nudges can effectively assist customers in identifying and choosing the most sustainable options available on the AFN online platforms.

Hopefully, this study can inspire future research on the intersection between online food provisioning services and sustainable food behaviours.

## References

Aprile, M. C., & Punzo, G. (2022). How environmental sustainability labels affect food choices: Assessing consumer preferences in southern Italy. *Journal of Cleaner Production, 332*, 130046. https://doi.org/10.1016/j.jclepro.2021.130046

Austgulen, M. H., Skuland, S. E., Schjøll, A., & Alfnes, F. (2018). Consumer readiness to reduce meat consumption for the purpose of environmental sustainability: Insights from Norway. *Sustainability, 10*(9), 3058. https://doi.org/10.3390/su10093058

Chen, M. F. (2020). Selecting environmental psychology theories to predict people's consumption intention of locally produced organic foods. *International Journal of Consumer Studies, 44*(5), 455–468. https://doi.org/10.1111/ijcs.12578

Corsi, A., & Novelli, S. (2018). Determinants of participation in AFNs and its value for consumers. In A. Corsi, F. Barbera, E. Dansero, & C. Peano

(Eds.), *Alternative Food Networks*. Palgrave Macmillan. https://doi.org/10.1007/978-3-319-90409-2_4

Dal Gobbo, A., Forno, F., & Magnani, N. (2022). Making "good food" more practicable? The reconfiguration of alternative food provisioning in the online world. *Sustainable Production and Consumption, 29*, 862–871. https://doi.org/10.1016/j.spc.2021.07.023

De Bernardi, P., Bertello, A., & Venuti, F. (2019). Online and on-site interactions within alternative food networks: Sustainability impact of knowledge-sharing practices. *Sustainability, 11*(5), 1457. https://doi.org/10.3390/su11051457

Eger, L., Komárková, L., Egerová, D., & Mičík, M. (2021). The effect of COVID-19 on consumer shopping behaviour: Generational cohort perspective. *Journal of Retailing and Consumer Services, 61*, 102542. https://doi.org/10.1016/j.jretconser.2021.102542

Filimonau, V., Beer, S., & Ermolaev, V. A. (2022). The COVID-19 pandemic and food consumption at home and away: An exploratory study of English households. *Socio-Economic Planning Sciences, 82*, 101125. https://doi.org/10.1016/j.seps.2021.101125

Forssell, S., & Lankoski, L. (2015). The sustainability promise of alternative food networks: An examination through "alternative" characteristics. *Agriculture and Human Values, 32*(1), 63–75. https://doi.org/10.1007/s10460-014-9516-4

Fuentes, C., Cegrell, O., & Vesterinen, J. (2021). Digitally enabling sustainable food shopping: App glitches, practice conflicts, and digital failure. *Journal of Retailing and Consumer Services, 61*, 102546. https://doi.org/10.1016/j.jretconser.2021.102546

Granheim, S. I., Løvhaug, A. L., Terragni, L., Torheim, L. E., & Thurston, M. (2022). Mapping the digital food environment: A systematic scoping review. *Obesity Reviews, 23*(1), e13356. https://doi.org/10.1111/obr.13356

Heidenstrøm, N., & Hebrok, M. (2021). Fridge studies–Rummage through the fridge to understand food waste. *Appetite, 165*, 105321. https://doi.org/10.1016/j.appet.2021.105321

Lu, M., Wang, R., & Li, P. (2022). Comparative analysis of online fresh food shopping behavior during normal and COVID-19 crisis periods. *British Food Journal, 124*(3), 968–986. https://doi.org/10.1108/BFJ-09-2020-0849

Lusk, J. L., & Briggeman, B. C. (2009). Food values. *American Journal of Agricultural Economics, 91*(1), 184–196. https://doi.org/10.1111/j.1467-8276.2008.01175.x

O'Neill, C., Hashem, S., Moran, C., & McCarthy, M. (2022). Thou shalt not waste: Unpacking consumption of local food. *Sustainable Production and Consumption, 29*, 851–861. https://doi.org/10.1016/j.spc.2021.06.016

O'Neill, C., McCarty, M., O'Reilly, S., & Alfnes, F. (2023). Food interests, preferences, and behaviours: A profile of the sustainable food consumer. *British Food Journal*. https://doi.org/10.1108/BFJ-09-2022-0762

Oncini, F., Bozzini, E., Forno, F., & Magnani, N. (2020). Towards food platforms? An analysis of online food provisioning services in Italy. *Geoforum, 114*, 172–180. https://doi.org/10.1016/j.geoforum.2020.06.004

Pascucci, S., Dentoni, D., Lombardi, A., & Cembalo, L. (2016). Sharing values or sharing costs? Understanding consumer participation in alternative food networks. *NJAS-Wageningen Journal of Life Sciences, 78*, 47–60. https://doi.org/10.1016/j.njas.2016.03.006

Preiss, P., Charão-Marques, F., & Wiskerke, J. S. (2017). Fostering sustainable urban-rural linkages through local food supply: A transnational analysis of collaborative food alliances. *Sustainability, 9*(7), 1155. https://doi.org/10.3390/su9071155

Samsioe, E., & Fuentes, C. (2022). Digitalizing shopping routines: Re-organizing household practices to enable sustainable food provisioning. *Sustainable Production and Consumption, 29*, 807–819. https://doi.org/10.1016/j.spc.2021.07.019

Shove, E. (2003). *Comfort, cleanliness and convenience: The social organization of normality*. Berg.

Stern, P. C., & Dietz, T. (1994). The value basis of environmental concern. *Journal of Social Issues, 50*(3), 65–84. https://doi.org/10.1111/j.1540-4560.1994.tb02420.x

Turner, B., & Hope, C. (2014). Ecological connections: Reimagining the role of farmers' markets. *Rural Society, 23*(2), 175–187. https://doi.org/10.5172/rsj.2014.23.2.175

van Bussel, L. M., Kuijsten, A., Mars, M., & van't Veer, P. (2022). Consumers' perceptions on food-related sustainability: A systematic review. *Journal of Cleaner Production, 130904*. https://doi.org/10.1016/j.jclepro.2022.130904

Verain, M. C., Bartels, J., Dagevos, H., Sijtsema, S. J., Onwezen, M. C., & Antonides, G. (2012). Segments of sustainable food consumers: A literature review. *International Journal of Consumer Studies, 36*(2), 123–132. https://doi.org/10.1111/j.1470-6431.2011.01082.x

Welch, D., & Warde, A. (2015). Theories of practice and sustainable consumption. In *Handbook of research on sustainable consumption*. Edward Elgar Publishing. https://doi.org/10.4337/9781783471270.00013

Ytreberg, N. S., Alfnes, F., & van Oort, B. (2023). Mapping of the digital climate nudges in Nordic online grocery stores. *Sustainable Production and Consumption, 37*, 202–212. https://doi.org/10.1016/j.spc.2023.02.018

# Driving the Digital and Sustainable Transition Through Law: Assessing the Food Consumer's Legal Toolkit

*Margherita Brunori*

## 1 Introduction

As the digital and platform economy is expanding and transforming society, the food sector is also affected by this change (Southerton & Fuentes, 2022; Samoggia et al., 2021; Rolandi, 2017). Whereas online food sales are a minority share in e-commerce, the Covid-19 pandemic has accelerated this transition and, since March 2020, online food sales have increased (OECD, 2020; Spoto, 2021). Besides the online version of supermarkets, several innovative online food provisioning services (OFPS) are blossoming, ranging from online speciality stores to farmers directly selling online; from food boxes home delivery to online platforms enabling producers and consumers to conclude contracts directly (Oncini et al., 2020, see also Oncini in this book).

M. Brunori (✉)
Department of Law, University of Brescia, Brescia, Italy
e-mail: margherita.brunori@unibs.it



A. Dulsrud, F. Forno (eds.), *Digital Food Provisioning in Times of Multiple Crises*, Consumption and Public Life,
https://doi.org/10.1007/978-3-031-46323-5_4

Regarding structural features, online food retailers are acquiring relevance in the food market, by setting rules and conditions for accessing the service, selecting the characteristics of food offered for sale, and establishing the contractual terms of transactions. Elements such as rankings, reviews, and information presentations have gained a strong reach in influencing people's choices (Lattanzi, 2022; Veltri et al., 2023). Regarding their contents, platforms are designed to facilitate the satisfaction of different consumers' demands—corresponding to specific people's needs or preferences—and they provide clients with particular typologies of food. Platforms might also enable new kinds of food relations by putting in contact, for instance, consumers with small farmers or artisanal food producers (Dal Gobbo et al., 2022).

Some of the dilemmas related to digital platforms are common to all categories of products, but food e-commerce should be, and indeed already is, given particular attention by the law, in consideration of its intrinsic characteristics. Food, unlike other commercial goods, is to be ingested regularly and therefore necessitates stronger control over its safety and more generally over its impact on people's health (Bolognini, 2017). In addition, food is aimed at satisfying a fundamental need recognized as the human right to adequate food, which calls—according to some strands of literature as well as social movements—for its exclusion from the category of commodities (Vivero-Pol et al., 2018).

If the human health–food nexus is an *acquis* for EU law, the sustainability–food one is also gaining a prominent place at the regulatory level. Food production, processing, distribution, consumption, and disposal are big contributors to CO2 emissions (ECA, 2021), environmental pollution, natural resources degradation (Westhoek et al., 2016) and, not least, social exploitation (Palumbo et al., 2022). Food systems, as other sectors of the economy, are therefore called to submit to the sustainability imperative and, arguably, drive the transition.

The law, in this picture, plays the role of ensuring that the market dynamics driven by the diffusion of online platforms do not hamper and instead facilitate the realization of objectives underpinning the legal system. As the situation evolves, preexisting regulatory settings have to pass the test of this new, evolving scenario.

In the EU context, food and e-commerce featured prominently in the Commission President von der Leyen's agenda (von der Leyen, 2020). On the one hand, the Commission's 2020 Communication "A Farm to Fork Strategy for a fair, healthy and environmentally-friendly food system" (hereafter, the Farm to Fork Strategy) (COM/2020/381) recognizes the agri-food as one of the key sectors needing to reach sustainability objectives. As food information will be increasingly called to leverage consumers' healthy and sustainable consumption choices, the Farm to Fork Strategy announced a reform of Regulation UE 1169/2011 on food information to consumers with the aim to "examine ways to harmonize voluntary green claims and to create a sustainable labelling framework that covers, in synergy with other relevant initiatives, the nutritional, climate, environmental and social aspects of food products" (Duncan et al., 2020). On 13 November 2020, the European Commission released its New Consumer Agenda, the new roadmap for consumer matters to guide the policy and regulatory activities from 2020 to 2025 (COM/2020/696). The document observes how the trend of buying locally, and online purchases of services and goods—including food—increased, and affirms the objective of empowering consumers to make informed choices and play an active role in the green and digital transition.

On the other hand, the European Commission Communication on the Digital Services Act Package affirmed that "digital services can support achieving Sustainable Development Goals by contributing to economic, social and environmental sustainability" (COM/2020/825).

Against this background, the objective of this chapter is to highlight the key elements of the EU regulatory framework surrounding OFPS, helping the nonspecialist reader understand to what extent the present legal discipline governing OFPS enables the consumer to make informed choices and play an active role in the green and digital transition. The chapter adopts a food consumer-centred perspective and interrogates the EU legal framework to clarify what are the instruments granted by the EU legislator to citizens for fulfilling their food-related rights, needs and choices while navigating the digital market.

## 2 Methodology and Structure of the Chapter

The chapter deploys mainly a legal research methodology: it assesses EU law by combining its analysis with case law and doctrinal interpretation of the norms, when relevant. However, as a way to illustrate the key emerging features and dilemmas of online food provisioning services, the first section will be dedicated to a synthetic appraisal of the structural characteristics of different OFPS and some of their contents, drawing from a list of Italian websites collected within the PLATEFORMS project's research activities. Against this picture, the chapter addresses applicable EU law from a consumer's perspective. It first gives an account of the evolution in EU food law, from guaranteeing food safety to the promotion of healthy diets. Secondly, it focuses on the growing attention by EU institutions to the protection of non-health food-related considerations. Thirdly, the analysis delves into emerging consumers' digital rights, i.e. consumer's rights concerning the digital environment. This recollection contributes to answering the initial research question in two ways. Firstly, it allows for clarifying how consumers' interests are considered in food law, what the instruments granted to citizens are in that regard, and what objectives such instruments are aimed at. Secondly, it explains how the information rights of food consumers are substantiated in the e-commerce environment. The chapter ends with a general reflection on the adequacy of the current regulatory framework in pursuing the aforementioned objectives. The scope of the chapter, following the PLATEFORMS project set-up, considers only online grocery shopping services; consequently, ready-to-eat meal delivery services lay outside of the main focus. The described legal framework, nevertheless, applies also to this kind of provisioning service.

## 3 Online Grocery Services: A Synthetic Overview of Structure, Contents, and Relevant Regulatory Frameworks

### *Exploring the Structure of Online Food Provisioning Services*

OFPS take several forms according to the structure, the role of the intermediary, and the profit or nonprofit aim. From the mapping made in the frame of the PLATEFORMS project, it emerged that they can be differentiated mainly into pipelines and platforms (see also Oncini in this book).

In the pipelines, the service provider manages all aspects of the transaction and sells the goods directly to the consumer. Some pipelines give more relevance to the suppliers' identity than others, but in that business arrangement, suppliers and consumers are mediated by the digital service provider. Under the European Union consumer law viewpoint, pipelines see a single food business operator as being responsible for the online sale and all information displayed on the website.

On the contrary, platforms are virtual spaces where consumers can choose from different suppliers. The difference in such platforms—which acquires relevance from the legal viewpoint—is that the sale is concluded directly between the supplier and the consumer, whereas the provider of the digital infrastructure, i.e. the online marketplace, operates as an intermediary between the two subjects. Virtually external to the sale, the platform provider can exert relevant influence on both consumers and business retailers. That fact, therefore, triggered the legislators' interest in an *ad hoc* treatment (COM/2017/229). The legislator recently considered this particular digital infrastructure because, in online marketplaces— differently from traditional e-commerce where the service provider sells products directly to the consumer—the platform provider has only an intermediary role. However, the intermediating subject indirectly enters into the contractual relationship by imposing certain contractual conditions that will regulate the transaction between the professional user of the online marketplace and the consumer. Consequently, in online marketplaces, consumers may be confused about the identity of the subject with whom they are concluding the contract.

EU law has, from several angles, updated the legal setting affecting this activity: it has reformed the regulatory framework on information society services with the Digital Services Act (Regulation EU 2022/2065); it has increased the responsibility of information society services providers as *gatekeepers* with the Digital Markets Acts (Regulation EU 2022/1925), and it has strengthened consumer rights in online marketplaces (Directive EU 2019/2161). The regulatory acts also have introduced legal definitions of the phenomenon. Directive EU 2019/2161 introduced the definition of *online marketplace* as "a service using software, including a website, part of a website or an application, operated by or on behalf of a trader which allows consumers to conclude distance contracts with other traders or consumers" (Directive EU 83/2011, art 2, n 17, as amended by Directive EU 2019/2161). The Digital Services Act defined an *online platform* as "a hosting service that, at the request of a recipient of the service, stores and disseminates information to the public [...]" (Regulation EU 2022/2065, art 3, lett i). The two definitions serve different purposes within the scope of the act that introduced them, meaning that an online food provisioning service can fit the definition of online platform under the Digital Services Act, and operate as an online marketplace under Directive EU 2019/2161. In legal terms, online platforms, or online marketplaces, are technically so only if they fit into the just-reported definitions.

### *Illustrating Some Relevant Content Displayed on Online Food Provisioning Services*

For assessing the contents of the OFPS, between June and November 2021, the author of this chapter conducted a web search on Italian food e-commerce websites collected from an existing database (Oncini et al., 2019), complemented with further search. Within each of the 28 selected websites, the author observed the general description of the service, the presence and characteristics of compulsory as well as voluntary information associated with single food items, and the Terms and Conditions section of the website. Extracts are translated from Italian by the author and by using Google Translator.

From this content search, it emerged that many of the surveyed OFPS specialized in food sales share relevant features. For instance, they often claim to shorten the food chain and connect food producers and consumers. Oftentimes a dedicated page narrates the identity and characteristics of the producers selling their products through the website. In addition, the website is designed in a way that enables consumers to choose from which producer their groceries can be purchased. This description of the service often echoes an attempt to connect the city with the countryside, to make fresh, genuine products available, and to support local or national family farmers by providing them with a wider market and by offering them fair prices. Another relevant characteristic is a description of the care used in selecting the supplier and products. Reviewed websites' contents greatly stress the high quality and sustainability criteria food producers must meet to be selected. Some claims extracted from the homepages of such websites are reported here as examples:

> Cortilia is an innovative e-commerce with a short supply chain: it selects for you the best agricultural and artisanal producers to do the shopping as if you were in the countryside. (Cortilia)
>
> Shop online directly from producers: find your Hive! (Alveare che dice sì)
>
> Seasonal fruit and vegetables with original shapes directly to your home with savings of up to 30% compared to the shop. (Babaco Market)
>
> Tasté a quick and easy way to have quality products from the farming world directly to your home or office. Our producers operate in compliance with the traditions of the local area and offer genuine and quality products. Tasté accompanies you in your shopping: from the choice of products to the table, to offer you tasty and healthy dishes. (Tasté)

> Il Contadino Online [The Online Peasants] was born from a long and consolidated experience in the wholesale export of fruit and vegetables—Made [only] in Italy—throughout Europe, combined with a passion for taste, healthiness and freshness. The idea of Il Contadino Online is to bring genuine and healthy products directly from production and comfortably to your home at very affordable prices. A real shopping revolution. An 'enough!' to fruit and vegetables and other beautiful, but not very good, food products from the traditional supermarket and a turning point towards products that are first of all HEALTHY and GOOD, even before being beautiful. (Il Contadino Online)

For these reasons, the provisioning services sharing such characteristics could appear as farmers e-markets, because they mimic the farmers market phenomenon, and try to replicate the positive externalities that farmers markets have (Pazzi, 2019), but using the digital tool. Beyond the similarities, however, it is plain to say that the offline and the online farmers markets are not the same things, because the digital doppelganger is a business-driven commercial activity. In online farmers markets the service operator reaps commercial benefits, such as capturing a percentage of sales revenue by exploiting the image of family farmers and the idea of proximity. In some cases, the platform operator's social responsibility is sustained by a good degree of transparency in the conditions applied, in others, the information is not so consistent. Regulatory instruments through which platforms and marketplaces demonstrate the claimed selection and sustainability vary. They can range from public schemes—organic agriculture certification, protected designation of origin or protected geographical indication, and the so-called benefit corporation format—to private instruments, such as internally developed ethical codes of conduct and the B Corp private certification (see more in Van Der Meulen, 2011).

The Table 1 below shows some examples of such sustainability-supporting instruments presented on the reviewed websites.

**Table 1** Online food provisioning services and their sustainability certifying instruments

| *Platform* | *Instrument* |
|---|---|
| Alveare che dice sì | Internally developed ethical code |
| Cortilia | Benefit corporation, with a dedicated sustainability standards webpage |
| Bioexpress | Organic certification |
| Local to You | B Corp private certification; organic certification |

In some cases, despite the claim(s) made, information is lacking altogether regarding the criteria used for selecting products or the producers. This lack of information may result in the violation of the norms against unfair commercial practices and consumers' rights, including food consumers' rights (Liuzzo et al., 2019; Lachenmeier et al., 2013).

Finally, great emphasis is placed on the specific qualities and characteristics of the products. Typically, food products are accompanied by a description celebrating their unique story, taste, and healthy attributes. The homepage often contains some text claiming the superior quality of the products compared to those of the conventional retail sector and the genuine food produced in harmony with nature by the small-scale producers. Another common feature is health claims associated with food products. This is an example:

> They are organic beans, they are rich in calcium, magnesium and potassium. Borlotti beans are an excellent source of protein for vegetarians. In addition, when well-cooked, they contain sulphur and chromium compounds that help keep blood sugar and cholesterol and triglyceride levels in the blood under control. So, they help prevent atherosclerosis and heart disease. (Retrieved from a webpage dedicated to a product description, in a section named "It's good also because…")

Consumers are particularly sensitive to food-related health claims, and their inadequate display can have adverse impacts on both protecting consumers' health and the market. Although health claims on food, as shown below, are carefully regulated by EU law, it is possible to observe that food e-commerce services are full of unauthorized health claims. They are especially frequent in online marketplaces, where the platform operates only as a hosting service and the food trader—often an artisanal food producer—has alone the responsibility to insert product information on the webpage. Unlawful or misleading claims are also more likely to be found where the website design is vague: for instance, where the product presentation webpage is not adequately structured to present required information, but instead presents only a general section dedicated to the description of the product, or, worse, a section encouraging the trader to add further information, for instance, in the mentioned product page section "It's good also because".

Adequate and transparent information is essential for the ability of consumers to make choices that are adequate to their needs and preferences and to be aware of their rights and risks when purchasing online (Purnhagen & Schebesta, 2019). This is why EU law set detailed requirements for information society services providers and traders who conclude distance contracts with consumers. In addition, food is considered by the legislator as worth a specific set of information requirements. As a consequence, the placement of food-offer to consumers through OFPS is subject to three main layers of norms: rules governing information society services and electronic commerce, consumer law, and food law (Costato & Albisinni, 2016; Bolognini, 2017).

From the consumer's perspective, those norms translate into information rights regarding, firstly, the website's general information that enables the user to easily identify and contact the responsible person; secondly, information concerning the contractual conditions, according to Directive EU 2011/83 on consumers' rights; and, thirdly, food information following Regulation EU 1169/2011 on food information to consumers, and other food labelling provisions, if pertinent.

Due to the specific requirements that food commerce finds in the EU legal framework, the next sections will first display the principles sustaining food information to consumers. Then, the emerging digital rights of consumers will be described.

## 4 Regulatory Trends in European Food Law

### *Protecting Consumers' Health*

The endeavour to work for "a highly competitive social market economy, aiming at full employment and social progress, and a high level of protection and improvement of the quality of the environment" is among the fundamental goals of the European Union (Treaty on European Union, art 3, par 3). Accordingly, the two principles underpinning food law—and any other matter of intervention of the EU—are the open market economy with free competition (Treaty on the Functioning of European Union, art 119) and the high standards in the matters of health, safety, environment, and consumer protection (Treaty on the Functioning of European Union, art 114).

Since the adoption of Regulation EC 178/2002, laying down the general principles and requirements of food law, the core rationale of EU food law is the commercialization of safe food in the single European market (Borghi et al., 2021). The initial consideration of the food–health nexus was portrayed mainly in terms of the absence of adverse substances for the human organism. With time, human health protection regarding food commercialization and consumption has progressively expanded, and particular legislation was set in response to a proliferation of market initiatives aimed at leveraging consumers' sensitivity to this aspect (McEldowney, 2020). In 2006, Regulation EC 1924/2006 was introduced to set conditions for the utilization of health claims in relation to food products (Strambi, 2016). Accordingly, only claims authorized by the European Food Safety Authority can be used and only in relation to the specific product for which they have been admitted. Later, with the adoption of Regulation EU 1169/2011 on food information to consumers, the European legislator deemed that full harmonization among national legislation was needed, therefore established the core principles of food information to consumers and issued a list of compulsory food information requirements (Paganizza, 2020).

In order to effectively protect consumers' choices, in consideration of the personal and fundamental interests that support and drive them, Regulation EU 1169/2011 sets three general conditions that need to be met in the labelling, the advertising, and the presentation of food. The first forbids information that is misleading about food characteristics, effects or properties and ingredients; the second posits that information should be "accurate, clear and easy to understand for the consumer"; the third affirms that information "shall not attribute to any food the property of preventing, treating or curing a human disease, nor refer to such properties" unless expressly permitted by Union law (Regulation EU 1169/2011, art 7). Regulation EU 1169/2011 introduced a dedicated provision to online food sales: information on food sold using distance communication should follow the same information requirements set by the Regulation and information should be available both before the purchase is concluded—except for expiry or best before date—and at the moment of delivery (art 14).

The most recent debate on food information relates to the so-called *front of pack* nutritional indication, which consists of a visual sign expressing the caloric content of processed products (Joint Research Centre, 2020). The evolution of the norms, as well as the Commission agenda expressed in the Farm to Fork Strategy, testifies to a progressive rise of institutional awareness of the close connection between diets and health, and the recognition that the abundance of highly processed food on the market has an impact in the surge of noncommunicable diseases in the European population.

In a nutshell, about food and health, European Union consumers have the right not only to buy food that is not directly harmful to health: the progressive elaboration of food consumers' rights is entailing the right to be meticulously informed on the food composition, quality, and use. Increasingly, food information is aimed at educating the consumer on how the specific product impacts their diets, and, soon, it might also inform about its environmental impact (Sirsi, 2011).

These provisions are crucial for the digital environment. Indeed, compared to the limited space of a label, the webpage or digital space gives much more room for traders to add voluntary information on the products they sell. In addition, consumers' perceptions of the nature and characteristics of the products can be biased by employing design elements and web architecture in addition to written elements (Di Lauro, 2018, 2020).

Regarding the enforcement of food law in the e-commerce context, Regulation EU 625/2017 on official controls widened the ground for operating official controls over food e-commerce and introduced three novelties. Firstly, it expressly required that operators shall provide the competent authorities with their name and legal form, and "the specific activities they carry out, including activities undertaken by means of distance communication, and the places under their control" (art 16). Secondly, the Regulation (art 36) enabled national authorities to "obtain samples through anonymously placed orders". Thirdly, it provided grounds to take measures to ensure compliance with the norms set by the Regulation, also through ordering "the cessation for an appropriate period of time of all or part of the activities of the operator concerned and, where relevant, of the internet sites it operates or employs" (art 138, par 2, lett i). However, as the sector is still expanding, it has to be ascertained whether the system for food control stands up to the challenge of ensuring the respect of food law in the digital environment (Plana Casado, 2021).

### *Food Law and the Cultural Dimension of Food*

Consumers' health is not the only aspect taken into consideration by the EU legislator. In the last decade, steps have been moved forward to reform also legislation on quality schemes for agricultural products and foodstuffs (Regulation EU 1151/2012), and on organic products (Regulation EU 848/2018). These reforms reflect the singularity of food as a commodity, and testimony to the attribution of value to immaterial features—although it has to be reminded that the principal driver for regulatory interventions on these matters is their economic relevance and power to shape and reshape food markets (Ambrosio et al., 2022).

Several of those aspects fall into the sphere of the cultural dimension of food. The acknowledgement of the significance of food traditions, for instance, entails the protection of methods expressing the immaterial cultural heritage of traditional food products; as a consequence, the processing of recognized traditional foods is exempted from the respect of certain standards (Paoloni, 2009; Strambi, 2010). Another relevant aspect is that of the origin of food. Expressing the deep connection of certain products with the territory in which they were developed, the origin is relevant not only for the material qualities that characterize the products obtained from a particular ecosystem and know-how but increasingly because consumers attach to it the satisfaction of a sense of identity (Bairati, 2020; Saija & Fabbio, 2020).

Conceptually connected to this aspect is the concept of proximity. Absorbing the narratives of territorial development, promotion of the local economy and sustainability, the reform of the common agricultural policy in 2013 introduced the concept of the *short food chain* (Canfora, 2016). Regulation EU 1305/2013 (art 35) on rural development offered financing opportunities for projects bringing together different actions and engaging in alternative forms of production or commercialization of agricultural products fostering short supply chains and local markets (Kapała, 2020). In the Regulation, a short supply chain is defined as a "supply chain involving a limited number of economic operators, committed to co-operation, local economic development, and close geographical and social relations between producers, processors and consumers" (art 2, lett m). This innovation is reflected in food law as well. Regulation EU 1151/2012 on quality schemes in fact asks the Commission to explore the feasibility of dedicated labelling for short food supply chains (see also COM/2013/855).

The breadth of functions that the indication of origin or provenance covers concerning food is well evidenced by a recent judgment of the European Union Court of Justice (CJEU Case C–363/18). In the case, the Court found that the indication of origin had the function of putting the consumers in the condition of developing their adequate food choices in consideration of ethical beliefs and, in particular, the respect of human rights. The case shows how ethical considerations find space in the application of food law and raises the question of to what extent food information should reach the satisfaction of consumers' interests (Bairati, 2020). The effects of the permeability of food law on ethics are reflected also in another case. A recent judgment of the European Union Court showed the tension between the respect of religious and ethical beliefs that the application of food law can cause when religious slaughtering procedures conflict with standards on animal wellbeing (CJEU Case C-336/19).

The cultural dimension of food could be summarized in the presence, in food law, of tools that can correctly indicate certain immaterial values enshrined in the connection of food with a specific tradition or territory. Recent cases showed how the boundaries of the function of food information are being pushed to encompass also considerations concerning ethical beliefs, including religion and respect for human rights (Di Lauro, 2021; Bairati, 2020).

### *Food Sustainability*

Between ethics and international commitments to the sustainable use of resources and action against climate change, environmental considerations in food law are also expanding. The new Regulation EU 848/2018 on organic products, for instance, improves the connection of the scheme with the common agricultural policy and shows deeper consideration of biodiversity, natural resources protection, the contribution of a nontoxic environment, and proximity (Lucifero, 2018).

Recent interventions of the EU on the matter speak also about the rising awareness of European institutions on the need to rethink the roots of the entire food system in the face of new challenges and raising sensibilities. In 2019, Regulation EU 178/2002 was enriched by a section enucleating the principle of risk communication, whereby the procedures supporting food safety risk analysis and the respective outcomes have to be duly justified, shared, and communicated, also by fostering the involvement of stakeholders in the process (Regulation EU 2019/1381). While the intervention was mainly promoted for safeguarding the legitimacy of the European Food Safety Authority and therefore ensuring continued trust in the European food market (Regulation EU 2019/1381, recitals 4 and 12), it showed how relevant the governance of food safety is for the wider public, and therefore how it requires bringing transparency and democratic principles at the core of the European food safety system

(Canfora, 2020). This reform stemmed from a *Refit* evaluation of general food law, in which the Commission acknowledged the increased awareness and susceptibility of food consumers to environmental and ethical considerations and admits certain limits of the current framework in this regard (SWD/2018/37).

For what concerns the sustainability of food and relative consumers' sensitivity, the match is still open. In the prosecution of this revision process, in 2020, a comprehensive reform agenda in the food and agriculture law was launched by the EU Commission. The already mentioned Farm to Fork Strategy addresses the sustainability challenge in the agriculture and food sectors, and while doing so it adopts for the first time a food systems approach. The Strategy recognizes the key role of the agri-food sector in achieving sustainability goals and affirms that food information will increasingly be called upon to leverage consumers' healthy and sustainable consumption choices. The Strategy acknowledges, on the one hand, a blossoming of private sustainability standards for filling the gaps left by public certification schemes concerning their scope (on the social impact, for instance, addressing the transparency and fairness of prices and working conditions), ambition, or reliability. On the other hand, it denounces that consumers are constantly exposed to sustainability claims which lack supportive data or justification (D'Addezio & Bolognini, 2021).

The general EU consumer law already offers some principles and guidance on how sustainability claims to commodities should be presented: the more emphasis is placed on product characteristics that may appeal to consumers' sensibilities, such as environmental or ethical considerations, the more accurate supporting information has to be. This aspect meets the provisions set out by Directive CE 2005/29 on Unfair Commercial Practices, as interpreted by the EU Commission guidelines on sustainability claims as well as for the application of the Directive in the context of digital commerce (COM/2016/320; SWD/2016/163; Multi-stakeholder Dialogue on Environmental Claims, 2016). However, a more incisive intervention is needed, especially in the food sector.

From this section, we have seen that food consumers' right to information is expanding, encompassing progressively new spheres. In principle, EU law already provides the tools for protecting immaterial consumers' interests such as the cultural and ethical dimensions of food. In practice, the increased market attention on such dimensions poses the norms, and their effective implementation, under stress. The upcoming reforms, put on the agenda by the EU Commission in the context of the Farm to Fork

Strategy, plan to further address this phenomenon, and it has to be seen what the consequences will be both on the promotion of sustainable food systems, and on the extent to which the dynamics that shape the digital food system will be considered.

## 5 Emerging Digital Rights for EU Food Consumers

### *Combining Different Responsibility Regimes*

Both Regulation EU 1169/2011 on food information to consumers and Regulation 178/2002 on the principles of food law aim at strengthening the protection of food consumers, therefore they extend the responsibility to ensure the respect of the food law requirements to all business operators in the food chain. The responsibility is shaped according to the effective role that the food business operator plays in the food chain, but the principle is that each food business operator has obligations under food law. Specifically, Regulation EU 1169/2011 (art 8, par 5) reads: "[…], food business operators, within the businesses under their control, shall ensure compliance with the requirements of food information law and relevant national provisions which are relevant to their activities and shall verify that such requirements are met."

The E-Commerce Directive CE 2000/31 has set common provisions for information society services. The main aim of the Directive was to set basic common rules and prevent Member States to impose conditions on national providers that would, in an unjustified manner, restrict the freedom of initiative of providers. The E-Commerce Directive exempted internet service providers from the liability which might derive from the information they merely conduct, and from the obligation to monitor the information they transmit or store (Directive 2000/31, art 15). This *safe harbour* approach was deemed necessary to promote the then-emerging internet market (Frosio, 2017, 2018). Nevertheless, it has to be said that by excluding the possibility for Member States to impose a general obligation to monitor, implicitly the norm allowed States to introduce monitoring obligations for specific fields (Directive 2000/31/EC, recital 47).

Since food law and the E-Commerce Directive have opposite approaches, how do operators' responsibilities have to be interpreted in the context of online food platforms? In what conditions the information society services provider can be captured under the category of food business operator and therefore be responsible for the respect of food law provisions? The EU Court of Justice contributed to clarifying the application of the liability exception, by distinguishing between the passive and the active role of online intermediaries: whereas in *mere conduit* and *caching* activities intermediaries are to be considered passive, in the *hosting* activities the Court found a more active role and a certain ability of the service provider to control over the content of the host. While performing hosting activities, therefore, online intermediaries have a stronger *duty of care* (Madiega, 2020).

For what concerns food directly, the Commission provided a partial answer to these interrogatives in a working document issued in 2015 on the application Regulation EU 1169/2011 (art 14) (Commission Services, 2015). This document, however unofficial, seemed to suggest that if an intermediary hosted the commercial webpage, then it would have the responsibility to ensure that the business user is aware of the need to comply with food law. This would pierce the mere conduit liability exception principle and attribute a specific responsibility to monitor, if not the compliance, at least the business users' awareness of food information law. This responsibility could be depicted as a specific monitoring requirement implicitly allowed by Directive CE 2000/31, as mentioned above.

### *The Digital Services Act Package*

The E-Commerce Directive was designed in a different environment than what society presently is. Information society and digital innovations profoundly reshaped people's economic behaviours online; the development of online platforms and the surfacing of the so-called platform economy posed new competition and fairness dilemmas (COM/2016/288). It is now evident how rapidly online intermediaries have acquired, thanks to a permissive regulatory framework, substantial economic powers (Graef, 2019; 2020/2019(INL); 2020/2018(INL); 2020/2022(INI)).

The Digital Services Act package tackles those emerging issues. Regulation EU 2022/2065 on Digital Services creates a comprehensive new design for information society services and addresses the functioning and impact of platforms, including *very big* platforms. A detailed description of the new framework goes beyond the scope of this chapter, but some elements are to be addressed as they have an impact on the emerging questions that online food provisioning platforms and marketplaces pose regarding the respect of consumers' rights.

First, the Regulation reaffirms the prohibition of general monitoring requirements. Nevertheless, it introduces the *good Samaritan clause*, which corrects the distorting effects of the liability exclusion as set by the E-Commerce Directive, which disincentivised voluntary monitoring carried out by online intermediaries (art 6). In particular, the liability exemption:

> shall not apply with respect to the liability under consumer protection law of online platforms that allow consumers to conclude distance contracts with traders, where such an online platform presents the specific item of information or otherwise enables the specific transaction at issue in a way that would lead an average consumer to believe that the information, or the product or service that is the object of the transaction, is provided either by the online platform itself or by a recipient of the service who is acting under its authority or control (art 6, par 3).

The Regulation, therefore, receipts the EU Court of Justice's interpretation of liability exception.

Second, it sets up a due diligence framework for service providers, which differentiates according to their typology and characteristics (Regulation EU 2022/2065, chapter III). As part of the introduced due diligence framework, a set of dedicated provisions apply to providers of online platforms allowing consumers to conclude distance contracts with traders: articles 30–32 are set up for ensuring the respect of consumers' rights in the digital environment, by introducing provisions for the traceability of traders, compliance by design, and right to information. The most relevant to the extent of this study is article 31 on *compliance by design*, which states that:

> [p]roviders of online platforms allowing consumers to conclude distance contracts with traders shall ensure that its online interface is designed and organised in a way that enables traders to comply with their obligations regarding pre-contractual information, compliance and product safety information under applicable Union law. (art 31, par 1)

In particular, the consumer should be able to easily retrieve information to identify the product, the trader, and, "where applicable, the information concerning the labelling and marking in compliance with rules of applicable Union law on product safety and product compliance" (art 31, par 2, lett c). The introduction of this article fills the normative gap or at least clears the ambiguity created by the previous regulatory setting: namely, the distance between the responsibility regime set for protecting consumers—and especially food consumers—rights, and the safe harbour of the 2000 E-Commerce Directive. By recalling the applicable labelling requirements, article 31 creates a bridge between the responsibility of food business operators regarding food information to consumers, as set by Regulation EU 1169/2011, and the responsibility of service providers of online platforms. The Digital Services Act in fact acknowledges the actual role and impact that the provider plays and balances its power with increased responsibilities towards the activities that are hosted on the platform. The new due diligence infrastructure suggests that the service provider has responsibility for food information according to Regulation EU 1169/2011 (art 8, par 5) and ensures compliance with the requirements of food information law, including by designing the website structure in a way that the provision of all required precontractual information is facilitated.

Not only the online service provider should create a compliance-enabling design of the platform, but it should also exert some degree of control on the respect of the norms:

> [p]roviders of online platforms allowing consumers to conclude distance contracts with traders shall make best efforts to assess whether such traders have provided the information referred to in paragraphs 1 and 2 before allowing them to offer their products or services on those platforms. After allowing the trader to offer products or services on its online platform that allows consumers to conclude distance contracts with traders, the provider shall make reasonable efforts to randomly check in any official, freely accessible and machine-readable online database or online interface whether the products or services offered have been identified as illegal. (Regulation EU 2022/2065, art 31, par 3)

The legislator here, however, decided not to grant an objective responsibility to the provider, but rather leave the judge the power to rule whether the provider's conduct reflects its "best efforts" or "reasonable efforts". How should this provision be read in conjunction with the responsibility regime set by Regulation UE 1169/2011 (art 8, par 5), which affirms that the food business operator shall verify that such requirements are met? Which regulatory framework prevails, and how to balance the different interests at stake, will have to be clarified by the Commission or by the EU Court of Justice.

### *New Information Rights: A Lot to Chew for Food Consumers (and Online Retailers)*

To address the fairness and transparency of the digital environment, and thus strengthen both the rights of consumers and the rights of the professionals using online e-commerce services, Regulation EU 2019/1150 introduced specific provisions strengthening the obligation of information service providers concerning the Terms and Conditions section of the website. The information service provider must also make clear who is responsible for the sale or supply and clarify applicable consumers' rights, and should also allow professional users of the service to easily access the service conditions, as they can be vulnerable to the power of the platform operator as well if the contractual conditions are not clear. The EU has also adopted a directive to enhance the transparency of certain mechanisms affecting consumers' choices. According to the new framework, marketplace operators have to publish the criteria determining the ranking, i.e. the order in which products appear, and ensure that reviews only come from people who have tried the product (Directive EU 2019/2161).

Recent EU law reforms are endeavouring to better attune consumer law and digital platforms governance. This results in an expansion of consumers' information rights and, more importantly, in a stronger responsibility of the digital service provider in guiding the professional users as well as the consumers towards a fairer and more transparent digital environment. The compliance-by-design provision, in particular, could have a direct impact on food e-commerce, by guiding traders to fulfil their responsibilities as food business operators, and consumers in raising awareness of their food-related rights. The general control duties established by the new Digital Services Act, in addition, could become a fruitful entry to leverage the provisions on official food controls set by Regulation EU 2017/625 on e-commerce channels and coordinate, in a preventive approach, the platform's internal control with the official one.

## 6 Conclusion

Drawing conclusions from the analysis of the legal framework interesting OFPS, the following key issues emerge. As innovation in the online food provisioning systems spurs, so does the food law field. Under the health, sustainability and ethics perspective, recent trends show how current boundaries of food consumer protection are stretched and pushed towards the goal of collocating food in a wider picture of food systems underpinned by the sustainability imperative. The respect of food information requirements, principles and fair information practices are widely challenged by online food sales, as consumers' rights and fair competition practices are often not respected on the internet. New instruments are put in place both in terms of responsibility and control, but their reach is yet to be tested *vis à vis* the fast-evolving reality of food e-commerce.

The resulting picture invites a careful reflection on the future of food law as, on the one hand, food information is increasingly called to leverage consumers' healthy and sustainable consumption choices and, on the other, online intermediaries become food law enforcers in the digital environment. From this perspective, the EU institutions must continue monitoring the challenges posed by OFPS in food consumption patterns and food information communication, and strengthen transparency and accountability. But mostly, they must facilitate and promote who contributes to sustainable food systems and ensure that they are not impaired by the advancement of OFPS—but rather supported and promoted by it.

## References


Ambrosio, L., Sajia, R., & Carrara, L. (2022). *Prodotti alimentari di qualità. Regole, casi, questioni*. CEDAM.

Bairati, L. (2020). L'etica del consumatore nel mercato globale. *Rivista di Diritto Alimentare, 4*, 21–36.

Bolognini, S. (2017). *Contrattazione a distanza e tutela del consumatore di prodotti alimentari*. Giuffré.

Borghi, P., Canfora, I., Di Lauro, A., & Russo, L. (Eds.). (2021). *Trattato di diritto alimentare Italiano e dell'Unione Europea*. Giuffrè Francis Lefebvre.

Canfora, I. (2016). Is the short food supply chain an efficient solution for sustainability in food market? *Agriculture and Agricultural Science Procedia, 8*, 402–407.

Canfora, I. (2020). L'evoluzione delle regole europee sulla trasparenza: verso un sistema di 'Sicurezza Alimentare 2.0'. *Rivista di Diritto Alimentare, XIV*(3), 4–14.

Costato, L., & Albisinni, F. (Eds.). (2016). *European and global food law*. CEDAM.

D'Addezio, M., & Bolognini, S. (2021). F-LAW. *Forestry law and food law, Approfondimenti su soggetti e strumenti giuridici all'insegna dei canoni di sostenibilità*. Giappichelli.

Dal Gobbo, A., Forno, F., & Magnani, N. (2022). Making "good food" more practicable? The reconfiguration of alternative food provisioning in the online world. *Sustainable Production and Consumption, 29*, 862–871.

Di Lauro, A. (2018). Neuroscienze e diritto. Ripensare il libero arbitrio in campo penale, civile e alimentare. *Rivista Italiana di Medicina Legale e del Diritto in Campo Sanitario, 4*, 1429–1450.

Di Lauro, A. (2020). Segni e territorio: quale co-design per quali modelli giuridici. *Rivista di Diritto Alimentare, 1*, 51–80.

Di Lauro, A. (2021). Religione del cibo: quale normatività per il benessere animale? *Rivista di Diritto Alimentare, 3*, 53–62.

Duncan, J., Rivera-Ferre, M., & Cleays, P. (2020). *The importance of food sovereignty for the Farm to Fork strategy and the New Green Deal: Insights and limits of the SAM and SAPEA reports.* Wageningen University & Research.

ECA. (2021). Common agricultural policy and climate: Half of EU climate spending but farm emissions are not decreasing. *Special Report, 16*, 2021. https://www.eca.europa.eu/en/Pages/DocItem.aspx?did=58913

Frosio, G. F. (2017). Reforming intermediary liability in the platform economy: A European digital single market strategy. *Northwestern U L Rev Online, 112*, 19–46.

Frosio, G. F. (2018). Why keep a dog and bark yourself? From intermediary liability to responsibility. *International Journal of Law and Information Technology, 26*(1), 1–33.

Joint Research Centre, European Commission, Storcksdieck genannt Bonsmann, S., Marandola, G., Ciriolo, E., et al. (2020). *Front-of-pack nutrition labelling scheme: A comprehensive review.* Publications Office. https://data.europa.eu/doi/10.2760/436998

Kapała, A. (2020). Legal status of direct sales of agricultural and food products in the legislation of selected EU member states. *Przegląd Prawa Rolnego, 26*(1), 65–77.

Lachenmeier, D. W., et al. (2013). Does European Union food policy privilege the internet market? Suggestions for a specialized regulatory framework. *Food Control, 30*(2), 705–713.

Lattanzi, A. (2022). "Average" consumers navigating the new digital food chain: Influencers, online reviews and rankings. *European Business Law Review, 33*(7), 1041–1056.

Liuzzo, G., Rolandi, S., Serraino, A., Piva, S., & Giacometti, F. (2019). Prepackaged food products business to consumer (B2C) distance selling and information obligations in Italian mass market retailers. *Italian Journal of Food Safety, 8*(3), 149–142.

Lucifero, N. (2018). Il regolamento (UE) 2018/848 sulla produzione biologica. Principi e regole del nuovo regime nel sistema del diritto agroalimentare europeo. *Rivista di Diritto Agrario, 3*, 477–508.

Madiega, T. (2020). Reform of the EU liability regime for online intermediaries: Background on the forthcoming digital services act *EPRS In-Depth analysis*, PE 649.404—May 2020, 2–3.

McEldowney, J. (2020). EU agricultural policy and health: Some historical and contemporary issues. *EPRS In-Depth analysis*, PE 659.302—October 2020, 1–34.

Multi-stakeholder Dialogue on Environmental Claims. (2016). Compliance Criteria on Environmental Claims, Multi-stakeholder advice to support the implementation/application of the Unfair Commercial Practices Directive 2005/29/EC, 1–15.

OECD. (2020). E-commerce in the time of COVID-19. *OECD Policy Responses to Coronavirus (COVID-19), 5.* https://www.oecd.org/coronavirus/policy-responses/e-commerce-in-the-time-of-covid-19-a2b78e8/#section-d1e662

Oncini, F., Bozzini, E., Forno, F., & Magnani, N. (2020). Towards food platforms? An analysis of online food provisioning services in Italy. *Geoforum, 144*, 172–180.

Oncini, F., Forno, F., Bozzini, E., Magnani, N., & Dal Gobbo, A. (2019). PLATEFORMS. Sustainable food platforms: Enabling sustainable food practices through sociotechnical innovation—WP1 mapping food provisioning services. https://uni.oslomet.no/plateforms/wp-content/uploads/sites/683/2020/04/2020407--PLATEFORMS-WP1_FINAL-report.pdf

Paganizza, V. (2020). A European overview on Regulation (EU) No 1169/2011 after the entry into force. *Rivista di Diritto Alimentare, 14*(1), 11–30.

Palumbo, L., Corrado, A., & Triandafyllidou, A. (2022). Migrant labour in the agri-food system in Europe: Unpacking the social and legal factors of exploitation. *European Journal of Migration and Law, 24*(2), 179–192.

Paoloni, L. (2009). Prodotti tradizionali e filiere produttive. *Rivista di Diritto Alimentare, 3*(1), 1–5.

Pazzi, U. (2019). I Mercati Contadini come esperienze di disintermediazione e di rivitalizzazione economica delle piccole aziende agricole. *Scienze del Territorio, 7*, 84–89.

Plana Casado, M. J. (2021). *E-food: Closing the online enforcement gap in the EU platform economy.* Springer.

Purnhagen, K., & Schebesta, H. (2019). Food labelling for consumers: EU law, regulation and policy options. *Policy Department for Citizens' Rights and Constitutional Affairs*, European Parliament, PE 608.871, 1–60.

Rolandi, S. (2017). Food e-commerce as a new tool for the growth of the economy. In M. Alabrese, M. Brunori, S. Rolandi, & A. Saba (Eds.), *Agricultural law: Current issues from a global perspective* (pp. 231–244). Springer.

Saija, R., & Fabbio, P. (Eds.). (2020). *La qualità e le qualità dei prodotti alimentari, tra regole e mercato.* Cedam.

Samoggia, A., Monticone, F., & Bertazzoli, A. (2021). Innovative digital technologies for purchasing and consumption in urban and regional agro-food systems: A systematic review. *Foods, 10*(2), 208.

Sirsi, E. (2011). Il diritto all'educazione del consumatore di alimenti. *Rivista di Diritto Agrario, 4*, 496–524.

Southerton, D., & Fuentes, C. (2022). Digital platforms and sustainable food consumption transitions. *Sustainable Production and Consumption, 29*, 805–806.

Spoto, G. (2021). Il mercato agroalimentare nell'era digitale innanzi all'emergenza covid-19. *Rivista di Diritto Alimentare, 1*, 54–63.

Strambi, G. (2010). I prodotti tradizionali e la politica di qualità dell'Unione Europea. *Rivista di Diritto Alimentare, 1*, 17–25.

Strambi, G. (2016). La disciplina europea sulle indicazioni nutrizionali e sulla salute: dieci anni di applicazione controversa. *Agricoltura Istituzioni Mercati, 1*, 28–52.

Van Der Meulen, B. (2011). *Private food law: Governing food chains through contract law, self-regulation, private standards, audits and certification schemes.* Wageningen Academic Publishers.

Veltri, G. A., Lupiáñez-Villanueva, F., Folkvord, F., Theben, A., & Gaskell, G. (2023). The impact of online platform transparency of information on consumers' choices. *Behavioural Public Policy, 7*(1), 55–82.

Vivero-Pol, J., Ferrando, T., De Schutter, O., & Mattei, U. (Eds.). (2018). *Routledge handbook of food as a commons.* Routledge.

von der Leyen, U. (2020). Political guidelines for the next European Commission 2019–2024, a union that strives for more—My agenda for Europe. Retrieved December 15, 2020, from www.ec.europa.eu/commission/sites/beta-political/files/political-guidelines-next-ommission_en.pdf

Westhoek, H., Ingram, J., Van Berkum, S., Özay, L., & Hajer, M. (2016). *Food systems and natural resources. A report of the working group on food systems of the international resource panel.* UNEP.

## Case Law, Legal Acts, and Institutional Documents

Directive (EC) 2000/31 of the European Parliament and of the Council of 8 June 2000 on certain legal aspects of information society services, in particular electronic commerce, in the internal market (Directive on electronic commerce) (OJ L 178, 17.7.2000).

Directive (EU) 2011/83 of the European Parliament and of the Council of 25 October 2011 on consumer rights (OJ L 304, 22.11.2011).

Directive (EU) 2019/2161 of the European Parliament and of the Council of 27 November 2019 amending Council Directive 93/13/EEC and Directives 98/6/EC, 2005/29/EC and 2011/83/EU of the European Parliament and of the Council as regards the better enforcement and modernisation of Union consumer protection rules (OJ L 328, 18.12.2019).

EU Court of Justice, Case C–336/19, 17 December 2020, *Centraal Israëlitisch Consistorie van België and Others.*

EU Court of Justice, Case C–363/18, 12 November 2019, *Organisation juive européenne, Vignoble Psagot Ltd v Ministre de l'Economie et des Finances.*

European Commission, Online platforms and the digital single market opportunities and challenges for Europe, 25.5.2016, COM (2016) 288 final.

European Commission Services, Issues relating to distance selling in the context of Regulation (EU) 1169/2011 on the provision of food information to consumers, 2015. Retrieved December 15, 2020, from www.mise.gov.it/images/stories/documenti/etichettatura_alimentare_vendita_%20distanza.pdf

European Commission, A comprehensive approach to stimulating cross-border e-commerce for Europe's citizens and businesses, COM (2016) 320 final.

European Commission, A farm to fork strategy for a fair, healthy and environmentally-friendly food system, COM (2020) 381 final, 13.

European Commission, Consumer health and food safety, COM (97) 183 final.

European Commission, Final report on the e-commerce sector inquiry, SWD (2017) 154 final, COM (2017) 229 final.

European Commission, New consumer agenda, strengthening consumer resilience for sustainable recovery, COM (2020) 696 final.

European Commission, Proposal for a Regulation of the European Parliament and of the Council on a single market for digital services (Digital Services Act) and amending Directive 2000/31/EC, COM (2020) 825 final.

European Commission, Staff working document on various aspects of short food supply chains accompanying the document "Report from the Commission to the European Parliament and the Council on the case for a local farming and direct sales labelling scheme", COM (2013) 866 final.

European Commission, Staff working document, Guidance on the implementation/application of Directive 2005/29/EC on unfair commercial practices accompanying the document "Communication from the Commission to the European Parliament, the Council, the European Economic and Social Committee and the Committee of the Regions—A comprehensive approach to stimulating cross-border e-commerce for Europe's citizens and businesses", SWD (2016) 163 final.

European Commission, Staff working document, The Refit evaluation of the General Food Law (Regulation (EC) No 178/2002), SWD (2018) 37 final.

European Parliament Resolution of 20 October 2020 on the Digital Services Act and fundamental rights issues posed (2020/2022(INI)).

European Parliament Resolution of 20 October 2020 with recommendations to the Commission on "A Digital Services Act: Adapting commercial and civil law rules for commercial entities operating online" (2020/2019(INL)).

European Parliament Resolution of 20 October 2020 with recommendations to the Commission on the "Digital Services Act: Improving the functioning of the single market" (2020/2018(INL)).

Regulation (EC) No 178/2002 of the European Parliament and of the Council of 28 January 2002 laying down the general principles and requirements of food law, establishing the European Food Safety Authority and laying down procedures in matters of food safety (OJ L 31, 1.2.2002).
Regulation (EC) No 1924/2006 of the European Parliament and of the Council of 20 December 2006 on nutrition and health claims made on foods (OJ L 404, 30.12.2006).
Regulation (EU) 1151/2012 of the European Parliament and of the Council of 21 November 2012 on quality schemes for agricultural products and foodstuffs (OJ L 343, 14.12.2012).
Regulation (EU) 1169/2011 of the European Parliament and of the Council of 25 October 2011 on the provision of food information to consumers (OJ L 304, 22.11.2011).
Regulation (EU) 2017/625 of the European Parliament and of the Council of 15 March 2017 on official controls and other official activities performed to ensure the application of food and feed law, rules on animal health and welfare, plant health and plant protection products (OJ L95, 7.04.2017).
Regulation (EU) 2018/848 of the European Parliament and of the Council of 30 May 2018 on organic production and labelling of organic products (OJ L 150, 14.6.2018).
Regulation (EU) 2019/1150 of the European Parliament and of the Council of 20 June 2019 on promoting fairness and transparency for business users of online intermediation services (OJ L 186, 11.7.2019).
Regulation (EU) 2019/1381 of the European Parliament and of the Council of 20 June 2019 on the transparency and sustainability of the EU risk assessment in the food chain (OJ L 231, 6.9.2019).
Regulation (EU) 2021/2115 of the European Parliament and of the Council of 2 December 2021 establishing rules on support for strategic plans to be drawn up by Member States under the common agricultural policy (CAP Strategic Plans) and financed by the European Agricultural Guarantee Fund (EAGF) and by the European Agricultural Fund for Rural Development (EAFRD) (OJ L 435, 6.12.2021).
Regulation (EU) 2022/1925 of the European Parliament and of the Council of 14 September 2022 on contestable and fair markets in the digital sector (Digital Markets Act) (OJ L265, 12.10.2022).
Regulation (EU) 2022/2065 of the European Parliament and of the Council of 19 October 2022 on a single market for digital services and amending Directive 2000/31/EC (Digital Services Act) (OJ L 277, 27.10.2022).

# Infrastructure, Impulsivity, and Waste. Exploring the (Un)sustainable Routines of Mainstream Food Shoppers

*Claire O'Neill, Shadi Hashem, Mary McCarthy, and Christopher Moran*

## 1 Introduction

Research suggests that grocery shopping plays an important role in addressing household food waste (Stefan et al., 2013; Stancu et al., 2016; Van Doorn, 2016). Its importance is attributed to multiple interrelated food purchasing routines that may cause discrepancies between what is purchased and what is consumed in the household. Most commonly, a

C. O'Neill (✉) • M. McCarthy • C. Moran
Department of Management and Marketing, University College Cork, Cork, Ireland
e-mail: claireoneill@ucc.ie; M.McCarthy@ucc.ie; c.w.moran@umail.ucc.ie

S. Hashem
Department of Agriculture Economics, Faculty of Agriculture - Cairo University, Cairo, Egypt
e-mail: shadi.hashem@cu.edu.eg



A. Dulsrud, F. Forno (eds.), *Digital Food Provisioning in Times of Multiple Crises*, Consumption and Public Life,
https://doi.org/10.1007/978-3-031-46323-5_5

lack of shopping planning (Stefan et al., 2013; Richter & Bokelmann, 2017), not checking food stocks at home (Schneider, 2008), shopping opportunistically (Ganglbauer et al., 2013), and not making a shopping list (Van Doorn, 2016) tend to result in overbuying, which oftentimes, unless managed effectively, leads to food waste (Farr-Wharton et al., 2014; Porpino et al., 2015; Richter & Bokelmann, 2017). Due to environmental, social, and economic challenges, shifting the current food system towards sustainability, from production to consumption, is a key objective for various policy goals at governmental and supranational levels (e.g., UN SDG 12) (United Nations, 2020). Grocery retail stores are being recognised as essential infrastructure that may enable sustainable shopping practices (Fuentes, 2014; Fuentes et al., 2019). This recognition is evidenced by increasingly providing sustainability-related product attributes (e.g., local, organic, fairtrade, free-range, etc.), by providing information on product origin, and by reducing packaging. However, grocery retailers may also present barriers to sustainable household consumption, indirectly, by influencing purchase decisions (Baker et al., 2009; Richter & Bokelmann, 2017)—for example, encouraging impulse buying through promotions and special offers (Zhang et al., 2007; Hultén & Vanyushyn, 2011; Porpino et al., 2015) which increases the likelihood consumers will buy more than is needed, potentially leading to household-level food waste (Evans, 2012; Godfray et al., 2010).

A critical issue in advancing global sustainable development around food consumption is tackling food waste. If we are to tackle end-of-line wastage, understanding the ways such waste arose is an important objective. To this end, this chapter aims to explore shopping routines of so-called mainstream shoppers (i.e., shoppers who frequent larger food retailers, both in-store and online, for their weekly food provision) to uncover ways their shopping practice contributes to food waste. Their shopping, storing, cooking, eating, and leftover storage practices all contribute in various ways to the disposal of foodstuffs. But for the purpose of this research, the shopping route through which they acquire most of their food is central. We explore, through the lens of social practice theory (SPT), ways shoppers shop and how stores' infrastructure (or online setting) and marketing cues affect what is placed in their basket and ultimately for use by the household. Therefore, our focus is on the first stage of the consumption process—food acquisition.

## 2 Literature Review

### *Shopping as Social Practice*

Food shopping is a highly integrated and routinised practice (Warde, 2005). Prior research contends that grocery shopping is a skilful, complex, and socially embedded practice (Meah & Watson, 2013; Fuentes & Svingstedt, 2017; Berg & Henriksson, 2020). Therefore, to effectively reduce unsustainable food shopping and enable sustainable food shopping, consumers may need to break old unsustainable routines to establish new sustainable ones (Fuentes, 2014). Practice theorists (Shove, 2010; Shove et al., 2012; Warde, 2014) argue the need to account for people's social, economic, and spatial conditions to better explain and understand consumption. The meanings (e.g., norms, ideals, and symbolism), competencies (e.g., skill and knowledge), and materials (e.g., objects, technologies, and infrastructure) shape the practice of food shopping. Food outlets present as material elements that require specific types of knowledge, skills, and understandings of what it means to 'do grocery shopping' (Dobernig & Schanes, 2019). Consumers encounter marketing activities designed to affect their shopping practice, with many feeling like victims of these marketing efforts when reasoning about surplus and food waste (Dobernig & Schanes, 2019). How meanings, competencies, and materialities are understood and managed can alter certain life changes or new life phases, such as having children, and thus new practices can emerge (Berg & Henriksson, 2020). Practices are not static, as Shove et al.'s (2012) dynamics of practice highlight. Practices evolve and recruit agency to enable transformation through various adjustments in everyday life.

It is emphasised that consumers need new meanings, competencies, and materiality to effectively break old routines and establish new ones (Shove et al., 2012). Akin to Shove et al.'s (2012) dynamics of practice, Fuentes et al. (2019) illustrated how reconfiguring shopping practices towards greater sustainability required new assemblages of meanings, competencies, and materialities in which the retail stores, as well as consumers' practices, had to be reorganised to support a new mode of sustainable shopping. They refer to this reorganisation as 'reskilling of the consumer' so that consumers break their old shopping routines and establish new ones (Fuentes et al., 2019).

Material infrastructures in-store and online play an important role in determining food acquisition. Examples of these in-store infrastructures include attractive displays (Mohan et al., 2013), store layout (Dash & Akshaya, 2016), promotional and in-store signage (Peck & Childers, 2006), in-store marketing (Dulsrud & Jacobsen, 2009), and store convenience (Mohan et al., 2013). Similarly, for the online context, adding a new material (i.e., website or app) holds several functions and meanings. Online food shopping is considered convenient and timesaving (Brashear et al., 2009), which holds high value for consumers with busy lifestyles (Berg & Henriksson, 2020). Several features associated with website infrastructure, such as quick-tap purchasing, user-friendliness of the website, the level of product information, navigability, and product display (Verhagen & van Dolen, 2011; Wells et al., 2011; Turkyilmaz et al., 2015), may influence shopping practice. Interestingly, Berg and Henriksson (2020) found that online food purchasing allowed for households to eat healthier, structure their eating habits and ultimately limit impulse buying so less food was purchased. However, other sustainability-related activities, such as reduced car use, were not evident amongst online food shoppers (Berg & Henriksson, 2020). On the other hand, Ilyuk (2018) attests that online shopping can increase consumers' psychological distance to food (i.e., less attachment to food because they do not physically connect to the items themselves and put them in a physical cart/basket) and thus may increase the tendency to waste that food.

### *Impulse Purchasing and Food Waste*

Impulse buying (IB) has received increased attention in consumption literature in recent years (Iyer et al., 2020; Amos et al., 2014) and has been identified as a significant antecedent of food waste (Lahath et al., 2021; Porpino et al., 2015; Schmidt, 2016). Impulse purchases can be classified into four categories (Stern, 1962): pure impulse (break purchasing pattern after seeing an encouraging item); reminder impulse (item remembered but not on the original shopping plan); suggestion impulse (appealing new product leads to purchase without any prior knowledge of the product); and planned impulse (unexpected saving in special offers and discounts). Any of these impulses may occur during shopping, depending on the context and presence of internal and external stimuli (Stern, 1962). Overall, IB can be defined as 'a sudden and immediate purchase with no pre-shopping intentions either to buy the specific product category or to fulfil a specific buying task' and 'occurs after experiencing an urge to buy and it tends to be spontaneous and without a lot of reflection' (Beatty &

Ferrell, 1998, p. 170). Whereas the concepts IB and unplanned purchases have been used interchangeably, it has been argued that they are distinct concepts. IB can be seen as a form of unplanned purchase although not all unplanned purchases are necessarily impulsive. This distinction is because unplanned purchases may occur due to the omission of a required product from a structured shopping list but may not be accompanied by an urge, or strong positive feelings (Amos et al., 2014). Nonetheless, consumers who shop without shopping lists and without prior planning become more exposed to external (e.g., marketing cues) and internal stimuli (e.g., emotions), which may trigger impulsive buying (Massara et al., 2014).

It has been asserted that IB is rooted in, and influenced by, internal stimuli, such as personality traits and emotional state (Iyer et al., 2020; Mohan et al., 2013; Muruganantham & Bhakat, 2013; Lahath et al., 2021; Verplanken & Herabadi, 2001), as well as motives (hedonic or utilitarian) and resources (such as time and money) (Amos et al., 2014; Iyer et al., 2020). External marketing stimuli, such as store infrastructure, merchandise, communications, and price promotions (Amos et al., 2014; Mohan et al., 2013), as well as social media (Lahath et al., 2021) have also been identified as determinants of IB. However, a recent meta-analytic review of IB literature (Iyer et al., 2020) has highlighted the context dependency of such determinants. Contextual variables that may moderate internal and external variables include shopping environment (e.g., online or in-store), product category (e.g., food or clothing), consumer decision stage, demographic characteristics, and whether the consumer is alone or accompanied whilst shopping (Iyer et al., 2020). Empirical studies have also suggested that neuroticism, which relates to 'negative affect, reactivity to stress, lack of the ability to control urges, and inability to delay gratification' can manifest in the form of IB (Olsen et al., 2016, p. 40). This point is further reinforced by Fenton-O'Creevy et al. (2018), who found that neurotic individuals can be driven to engage in IB in order to relieve negative emotions through instant gratification—a strategy for mood regulation. Indeed, recent survey data have indicated that online shoppers miss the 'feel good', spontaneous aspect of impulse buying when shopping online compared to in-store (Farber, 2020). As well as internal factors, external stimuli, such as in-store promotions and shelf placement (Iyer et al., 2020) and store environment (Mohan et al., 2013), have been identified as catalysts of IB. However, in their study exploring the influence of promotional techniques on IB, such as expiration date-based pricing (EDBP), Yassin and Soares (2021) found that EDBP does not significantly affect IB.

In the context of addressing food waste, there is a call for a shift away from sales promotions of products nearing their expiration date by adopting strategies centred on environmental motivations (Yassin & Soares, 2021). Recent quantitative research looking at food consumption has indicated that neuroticism and social media usage were strongly associated with IB and subsequently food waste; impulse buying plays a mediating role between neuroticism and food waste and between social media usage and food waste (Lahath et al., 2021). Further empirical research on grocery shopping (Bellini et al., 2017) found that preparation in advance of shopping reduced impulsive purchases, a finding which supports findings of previous work (see Dholakia, 2000). Personality characteristics, notably the impulse-buying tendency and the shopping-enjoyment tendency, significantly influence impulse buying in grocery retail settings (Bellini et al., 2017; Mohan et al., 2013). Importantly, in the context of sustainability in the food domain, excessive purchasing, of which impulse buying is a significant antecedent, has been identified as a notable contributor to household food waste (Porpino et al., 2015) and thus must be addressed by both retailers and consumers.

## 3 Method

Social practice theory was used as the methodological approach to study the food acquisition practices of two types of mainstream food shoppers—those who shop in-store and those who shop online. An ethnographic approach using in-depth interviews and demonstration techniques, such as photo diaries, nonparticipant observations, kitchen walkabouts, and digital walkthroughs, were used. All interviews were conducted in participants' homes. The rich data set allows for triangulation of the data (Denzin, 1970). There were 21 (6 of whom were online shoppers) participants in total, 19 of whom were female and 2 of whom were male. Although the online shoppers represent a rather small portion of the sample, the data for this group provided an interesting account of the online food shoppers' experiences. Furthermore, the accounts provided displayed a consistency that differed from the consistency in accounts of in-store shoppers in some interesting ways. Convenience sampling using two degrees of separation from the researcher was used, an approach which fitted the predefined sampling criteria of the PLATEFORMS project. Participants were those who identified as the primary or co-primary grocery shopper for their household; thus, gender balance was not sought, nor achieved. Interviews took place between September 2019 and February 2020. The profile of participants is detailed in Table 1 below.

**Table 1** Profile of participants

| *Interview number* | *Channel* | *Pseudonym* | *Gender* | *Age* | *Social class*[a] | *Household composition* |
|---|---|---|---|---|---|---|
| Int. 1 | In-store | Theo | M | 20–29 | Managerial and technical | Lives with partner |
| Int. 2 | In-store | Ella | F | 40–49 | Managerial and technical | Family of 5 |
| Int. 3 | In-store | David | M | 60–69 | Student | Lives alone |
| Int. 4 | In-store | Maggie | F | 40–49 | Managerial and technical | Family with 3 children |
| Int. 5 | In-store | Sara | F | 40–49 | Managerial and technical | Family of 4 |
| Int. 6 | In-store | Jackie | F | 30–39 | Managerial and technical | Lives with partner |
| Int. 7 | In-store | Lily | F | 60–69 | Semiskilled | Family of 4 |
| Int. 8 | In-store | Abi | F | 40–49 | Professional Workers | Family with 2 children |
| Int. 9 | In-store | Ber | F | 30–39 | Semiskilled | Family with 2 children |
| Int. 10 | In-store | Teresa | F | 50–59 | Semiskilled | Family of 4 |
| Int. 11 | In-store | Kate | F | 30–39 | Skilled manual | Family with 1 child |
| Int. 12 | In-store | Molly | F | 40–49 | Managerial and technical | Family with 4 children |
| Int. 13 | In-store | Naomi | F | 20–29 | Student | Lives alone |
| Int. 14 | In-store | Chloe | F | 40–49 | Managerial and technical | Family with 2 children |
| Int. 15 | In-store | Dawn | F | 30–39 | Professional workers | Lives with partner |
| Int. 16 | Online | Eva | F | 30–39 | Managerial and technical | Family with 2 children |
| Int. 17 | Online | Darcie | F | 30–39 | Managerial and technical | Lives with partner |
| Int. 18 | Online | Cindy | F | 30–39 | Managerial and technical | Family with 3 children |
| Int. 19 | Online | Melissa | F | 40–49 | Managerial and technical | Family with 2 children |
| Int. 20 | Online | Alyssa | F | 50–59 | Non-manual | Family of 3 |
| Int. 21 | Online | Camryn | F | 30–39 | Professional Workers | Family with 1 child |

[a]Based on the Central Statistics Office's (CSO) social class groups (introduced in 1996), which are defined on the basis of occupation: professional workers; managerial and technical; non-manual; skilled manual; semiskilled; unskilled; all others gainfully occupied; and unknown

The two-hour interviews, including the walkabout, were audio-recorded. Additionally, the navigation of the online shopping processes and participant reflections on it were video-recorded. Braun and Clarke's (2006) thematic analysis framework was employed to identify emerging themes. This analysis was informed by SPT to make sense of the identified patterns.[1] Themes identified in the interview transcripts were compared with those in the other data sources (photo diary and nonparticipant observation photos and videos from the kitchen and digital walkabouts), a comparison which allows for further confidence in the findings. Ethical approval was granted prior to commencing the research.

## 4 Findings

Food acquisition routines play an important role in consumption practices in the home. Shoppers were identified as either organised or disorganised in their food shopping, characteristics which had consequences for their food consumption. Food in most homes was centred around the needs of the family and healthy eating ideals. Most shoppers, irrespective of income, were concerned with budget, with their main focus on value for money via special offers. Their tendency towards seeking special offers oftentimes resulted in impulse purchases that were usually not integrated into their routine food consumption. The findings are presented here in three themes: First, two basic profiles of shoppers are presented, one representing an organised shopping style and another representing a disorganised shopping style. Second, consideration is given to the context within which decisions around food are made, focusing on routines in transition, where changing roles and ideals around family needs and healthy eating aspirations emerged. Third, shopping infrastructure and impulse purchasing of special offers present as having notable implications for waste (both food and financial) and for the ability of households to integrate an overprovisioning of food into their food consumption routines. The findings are based on how the meanings, materials, and competencies of these shopping routines are interlinked, resulting in (un)sustainable consumption.

[1] Data were analysed using the NVivo 12 software tool for qualitative data analysis. Coding reliability was strengthened by building consensus among the research team via group discussion to check the reliability of data themes and ensure interpretive coherency.

### *Food Shopper Styles*

Mainstream food shoppers can be classified into two broad groups—organised shoppers and disorganised shoppers. Food acquisition for organised shoppers is highly routinised and structured as a big weekly shopping trip, either in-store or online. Food shopping tends to be completed on the same day every week, commonly at the weekend or early in the week. This shopping style is characterised by its highly structured and routinised nature with shoppers engaging in shopping lists, meal planning, stock checking, budget control, and only engaging in planned impulses, although almost all households engage in some top-up shopping, usually for fresh vegetables, milk, and/or bread. Organised shoppers frequent in-store supermarkets that are simplistic in their layout, facilitating less browsing and impulsivity, or shop online, a shopping style which suits their busy lifestyles. On the other hand, food acquisition for disorganised food shoppers is unstructured and sporadic. These shoppers tend to be in-store shoppers. However, they may also be online shoppers who rely on the online platform to organise their shopping via features such as past orders or usual items lists, but are otherwise disorganised in their approach to food shopping. These shoppers appear to be reactive to store environments, engaging in impulsive food purchases. Disorganised shoppers do not meal plan nor write shopping lists but rather shop frequently to satisfy needs and moods. There is a disorganisation in their shopping style but they are flexible in their approach to food acquisition and consumption. This style of shopper is heavily influenced by special offers and engages in both pure and planned impulses, behaviour which has consequences for sustainability in terms of overprovisioning of food and/or food waste.

For organised shoppers, meal planning and checking food stocks were key elements in creating shopping lists before going shopping. Theo explains:

> The way we do it is we start in the fridge here and write down on a piece of paper and go from there. You kind of see what we need for the home first and what the plan for the week was and how much food we needed, then we would work around that. (int.1, M, In-store shopper)

The family activities and plan for the week were central in determining how food shopping was organised and integrated into family life. Molly discussed how her husband organises a shopping plan as being determined by their availability at home, the time frame available for cooking in the evening, and family preferences. Their ability to plan their weekly meals accurately and to plan what food was needed was central to their organised shopping and household consumption routines.

> … Dominic did it up, so he did a fortnightly plan, and he very much would sit down and look at it. It's also dependent on the evening; who's home, what timeframe they have for whatever's put together. So, that's one of the first things, and then it literally is writing down … you just have your list of, so everybody likes vegetarian Bolognese, everybody likes omelette and chips, […] that one's easy for Monday, that one's easy for Wednesday, that one's easy, and we leave Sunday wide open. So then, again, looking at that and then pulling your shopping list back off that. (int. 13, In-store shopper)

In contrast, disorganised shoppers are spontaneous and are unable to plan their shopping properly, which justifies their frequent shopping practices. This is illustrated in Chloe's account of her household's shopping practices "… we're not very good planners. We are very spontaneous eaters as well. It's whatever you feel like, we'll go shopping" (int. 14, In-store shopper).

The shoppers who presented as flexible and spontaneous enjoyed the frequency of their grocery shopping because it gratified their daily food desires. For Maggie, mood played an important role in dictating everyday food practices, with food shopping depending on instant meal planning on the day:

> … I can't really think of the next dinner; it depends on my mood. I'm only ever thinking about two days ahead, and then Tuesday or Wednesday I would probably be going to the butcher's or [supermarket] to get my meat for that night's dinner. So yes, I don't really plan, I've no list. (int. 4, In-store shopper)

This lack of planning in terms of both grocery needs and meal planning resulted in an increased opportunity for impulse buying. Chloe admits to browsing all the shop aisles, leading to impulsive behaviour to take advantage of special offers and purchase what appeals to her *"so in [supermarket name] I go down every single aisle, I get sucked into some of the produce that they're selling on a special offer"* (int. 14, In-store shopper).

Interestingly, the online environment presents a hybrid of an organised and disorganised shopper. The sliding door of online shopping is such that the list is already there waiting for the shopper to alter to their needs.

Shopping lists for online shoppers were usually composed in real time, from previous orders or usuals or favourites lists, rather than preplanned. The outcome in terms of the final list is organised and comprises all the necessities to feed the family for the week, with full dishes/meals being a priority, but the shopper's effort is minimal. Melissa gave an enthusiastic account of this process:

> You have a favourite list when you're doing your shopping online, so I use that list more than I would deviate. I don't really deviate off that list. It gets a bit boring, we have the same kind of stuff all the time, you don't have any variety when you do it online all the time [...] when I do it online I can do it at 7 o'clock in the evening for the next day, I can sit down and think about it and just go, 'click, click, click.' I can check my cupboards if I think, 'Do I need that?' I can go and check. So, I am here doing it. That way is more convenient for me. (int. 19, Online shopper)

Although acknowledging that relying on the same virtual list of favourites every week results in food shopping lacking variety or being boring, a sentiment shared by others, the convenience and ease of shopping trumps variety. The infrastructure of the online app or website ensures that shoppers are organised (and nonimpulsive) but this organisation is facilitated by the website itself rather than by a trait of the shopper. Additionally, for online shoppers, meal planning was not preplanned, but rather occurred in real time, as Eva describes:

> I can't imagine myself being the kind of person who would sit down in the traditional manner and write out a list [...] the online shopping, I think, makes it a lot more organised. I know my salmon is going to go with my rice, my steak can go with mashed potatoes or something like that. And then I pick up items that might make a fry or something for the weekend. (int. 16, Online shopper)

The online platform and the nature of the virtual basket at the end before checkout enables shoppers to meal plan on the spot and return to shopping if they need to add anything extra to make a meal. Many online shoppers began shopping this way because of family demands or because of the ad hoc or sporadic nature of their food shopping when they used to shop in-store.

In the case of online shoppers, the material infrastructure of the app reorganises the shopping practice to one that is more structured, thus allowing the shopper to continue to be unprepared but to achieve a balanced food shopping outcome—the material infrastructure of the app's features puts structure around online shoppers' otherwise unpreparedness for food shopping.

### *Routines in Transition*

Food acquisition routines appear to be transitioning to align to new food ideals and roles (e.g., the caring, responsible parent). Family care, that is, ensuring there was enough food for family meals, particularly for young families or those in the formation stage, and ideals around healthy eating, dominated the ways in which shoppers planned shopping and acquired food. Having the available ingredients for family meals determined how and where shoppers acquired food, with old routines being abandoned in favour of new ones that facilitated a more organised approach to cooking and eating. However, transitioning to healthy eating and health aspirations for both one's self and the family oftentimes resulted in overprovisioning of healthy options that ultimately ended up as food waste. Here we see new contexts emerging around family togetherness and ideals of healthy food provisioning.

Transitions to new ways of doing were most evident in the online shoppers who were once sporadic in their food acquisition routines before switching to online shopping. Darcie used to rely on a small shop that had only a limited variety of produce and she described their previous shopping style as imbalanced and disorganised due to the limited variety of produce they had access to. This deficiency caused them to make frequent trips to shops during the week, which increased their expenditure on grocery shopping,

> We didn't really have a balanced weekly shop … there were parts of a dish. […] So, I suppose it was to try to have more balanced eating and eat a bit better and also to try to save a bit in terms of how much food we buy and the type of food we buy. (int. 17, Online shopper)

However, a household's desires to be a good provider of food for the family and to ensure adequate food availability in the home were reported as causes of increasing food waste by some shoppers. A frequent example was the conundrum of their desire to provide healthy options for children while also trying to avoid food waste, as illustrated by Eva:

> So I tend to overbuy to make sure there's stuff in the house and we never make it through all of it ... we tend to get through all the meats. Then we don't tend to get through all the fruit and veg, so then at the end of the week, before the new shop comes in, I have a quick look and there's generally half a pack of tomatoes, half a broccoli, a load of mouldy apples or whatever that all get shoved in the bin [...]. (int. 16, Online shopper)

Here, there is a routine of overpurchasing healthy food on a weekly basis for the family as a means of ensuring that ideals around sufficient stocks of food and healthy diets have the best possibility of being met:

> I'm a bit paranoid about there always being food in the house, so I'll over order constantly. And I have more guilt about there not being a lot of food in the house for the kids than I do about the whole food waste thing, even though I know it's bad. (int. 16, Online shopper)

Providing healthy options for her family is prioritised over potential food waste. There is a sense of inevitability regarding food waste, where households lack the competencies to reimagine different uses for unwanted food or the capacity to store or preserve food for future use.

Healthy eating aspirations for one's self were cited too by shoppers in their routines of purchasing fresh fruit and vegetables, along with ensuring there were healthy ingredients available for meals. The recollection of Darcie, who used to be disorganised and sporadic in her in-store shopping, illustrates how purchasing food via the online app results in more fresh fruit and vegetables being purchased, but not necessarily routinely eaten:

> Sometimes we might be a bit overambitious as well with what we order (online) [...] I would probably be less inclined to pick up fruit (in-store) because I'd be like I'm just going in to get stuff for dinner or something like that. Whereas if I did an online shop it was more likely to be there (on the list). There was a good chance that it wouldn't all be eaten but at the same time there was more of a chance that we would snack on it because it's there. It's hard to find that balance. (int. 17, Online shopper)

Where the online app functions as an easy reminder to purchase healthy food (via the favourites list) there was an overambition to be healthy, which resulted in some minor increase in healthy eating but unfortunately also resulted in increased food waste.

### *Infrastructure and Impulsivity*

Shop infrastructure, such as size, store layout/structure, and presence of promotions and special offers are important material elements in accounting for households' impulse purchases. Several households expressed that the size and layout of the shop was central to the time spent shopping, where their shopping at smaller shops is described as 'quicker', 'functional', and 'no-nonsense', compared to their shopping at larger stores where they invested more time. Online supermarkets' website infrastructure and service features were essential material elements in promoting household organisation and planning competencies in online grocery shopping practices. Service features, such as the product search bar, saved lists of previous orders and list of favourites, facilitated the automatic repetition of previous orders. For in-store shoppers, store infrastructure in terms of layout and where special offers were located were subject to households' pure impulse purchases. This is highlighted by Dawn:

> … you have much more targeted marketing in [supermarket name]. You walk in and you smell … you have to get through all the special offers at the front, the quiches, and then you have to get through the lovely soups and there's a hot food counter on the left there as you go past the bread. […] I find I tend to pick up more stuff that I didn't really need. (int. 15, In-store shopper)

However, in avoiding store browsing as part of a purposeful strategy among organised shoppers, importantly, shopper competence such as store familiarity was central in carrying out efficient shopping and avoiding impulse buying, as Theo illustrates:

> When you go to a supermarket, you always know where things are, you go in, you whiz through it in 40 minutes, and you are out … because we go in with what we want. We don't really venture. (int. 1, In-store shopper)

Impulse buying is more commonly practised by shoppers who, despite organising their weekly shopping trips, tend to divert from the shopping plan and written lists when faced with a special offer, or indeed are disorganised in their overall shopping style and are thus open to impulse buying. An otherwise organised shopper who deviates from her list when met with special offers is Jackie:

> … more than likely if there's a particular brand I like and if they're on offer I'll pick them up even if they're not on my list. So, the list is definitely not something I'm held by, for sure … (int. 6, In-store shopper)

Diversion from their shopping plan is often linked to impulse purchases upon encountering a new food item (pure impulse) or encountering items they remember they need to buy (reminder impulse), as Abi noted:

> That could be just a packet of biscuits. It could be a melon that's not on the list. I will just say, 'The melons look nice. I will pick up a melon.' It could be I didn't write down the smoked salmon or I will see something I think the kids might like and I will pick it up. Or I might have forgotten to put diluted orange on it [the list] or diluted blackcurrant and I will say, 'I have to get that.' (int. 8, In-store shopper)

Shop infrastructure in terms of store layout and size was a functional matter in households' impulsive shopping practices by increasing time spent in shops and influencing navigation practices as illustrated by Jackie:

> If I go to [supermarket, simplistic layout] I'd say it takes me 15 minutes, 20 minutes, even if I were to do a full shop. Because [it] is so accessible I suppose, it's four aisles maybe that I go up and down, whereas [large supermarket] I probably go up and down 12 aisles, so I just go in and I just have the pattern of just go down and around the fruit and veg, then down by the bakery and over by the butcher, fish counter and just up and down, then along the aisles … sometimes I'm like oh god, do I need to go down this aisle or not. It really is quite big … But I do get distracted by things or the whole cheese section and the olives; I just think I wander a little bit too. (int. 6, In-store shopper)

In contrast, as the previous section suggests, online shopping was commonly utilised by shoppers as enabling shopping efficiency and avoiding impulse purchases. There was as a result of reduced temptations and distractions normally associated with in-store shopping for Eva "… the action of scrolling through an app where you have to go into categories and subcategories kind of focusses my attention" (int. 16, Online shopper).

Her online shopping is structured and organised by the layout of the app. Sticking to online shopping lists' the usuals tab enables households to prevent deviations that often lead them to buy too much or make unnecessary food purchases. This is further illustrated by Cindy, who avoids impulse purchases via online shopping, which ultimately helps her reduce her overall food expenditures and avoid unhealthy foods.

> I do it online because I'm saving. You're not buying—as you go into a shop, picking up stuff that, 'Shall we try this as a new thing? We'll try this,' because you're not seeing it, you're not physically seeing it. [...] I guess if I was in-store, that's going to probably change because you are looking at a big shelf of biscuits as opposed to just ... I'm not going to bother browsing through pages online of biscuits, that's just not really the same thing. (int. 18, Online shopper)

However, online shoppers are not completely shielded from impulse buying. Online supermarket websites/apps display special-offer suggestions for shoppers when they view their shopping baskets, a feature which ultimately promotes planned impulses. For example, Alyssa revealed she would always go back to her shopping basket to check if there were special offers and potential savings she missed on products she selected:

> Once I go into my usuals, into my basket, and I just go through it and see, and then it might say, 'with savings, you missed out, and you could have got three for the ...' you know, this sort of thing, and then you go back. And that would be one now where you would say you'll get two for €5 or three for €5 and you'd be, 'would you know, I missed out on that?' and then you click it then. And then at the very end it says what you saved on. (int. 20, Online shopper)

Here, shoppers are overprovisioning on one or more food items, as prompted by the app, that they had originally planned to buy less of. This impulse is done based on a special offer but without consideration being given to its integration into the weekly food cycle.

Special offers on perishable fruits and vegetables were particularly attractive to households that desired to be healthier, but these were purchased with no specific plan around when they would be consumed. Thus, an abstract goal to eat/be healthier leads to an unplanned purchase where no related consumption plan is in place and therefore has implications for food waste, as Ella illustrates:

> Fruit is one of the things that goes off the fastest for me, or the thing that gets wasted the most actually in the house. [...] One of the things that would always reel me in is if it's like let's say the bananas I bought today are two bunches for €3. Or the organic apples, it's like 2 packets for a fiver [...] I think you're only saving about 25 cents or something like that, but I'll still

> always make sure that there's two of the two for three in the trolley and there's two of the two for five, which is ridiculous. I threw out pears this morning and I threw out a mouldy lemon. I think I threw out a nectarine as well. (int. 2, In-store shopper)

Ultimately, special offers, although also tied to a sense of achieving a second goal of getting value for money, encouraged shoppers to purchase perishable foods on impulse without a clear plan for integrating that extra food into the household's food consumption routines and thus resulted in food (and financial) waste. At the moment of purchase, the dominant influence is the feeling/view that the special-offer foods will be consumed and thereby help support a healthier lifestyle in the household. This feeling appears stronger than the potential of wasting the extra food, and thus the loss of money is experienced later. However, the cycle is not broken week-to-week, because healthy eating aspirations continue. However, some food routinely bought on impulse was either consumed immediately during the week or instead frozen for later use, particularly among organised shoppers, such as Theo, who described himself as a 'stickler' for not wasting food, despite often making planned impulse purchases based on special offers on fruits and meats. His organised household routines meant that he would divide meat bought on planned impulse, cook half, and freeze the rest for later use:

> We only freeze stuff if it was in large quantities like eight chicken breasts or if it was like ... two stacks of mince. It might be if you bought a deal of mince or something like that or chicken breasts, you would maybe freeze half of it, but it would still be used in the week. (int. 1, In-store shopper)

Success stories regarding special offers were where shoppers had particular meal planning strategies to integrate impulse items into their consumption routines. For example, Darcie and her husband would tailor their weekly meal plan with special offers seen on the online supermarket webpage to ensure efficient use of these offers when they shopped online:

> We would look at special offers online, on the meat or the fish, and then go okay, there's stewing lamb so we'll have a stew or there's fish; we'll have fish pie, or whatever. (int. 17, Online shopper)

In their last online shopping trip, Darcie's husband found a pork loin on offer,

> … that's why I got that pork loin. It was on offer and I was like 'oh, what do you do with pork loin? …' So, I just opened up another tab and I went on the [popular food recipe website], as in what do you have with pork loin? (int. 17, Online shopper)

Indeed, shopping online was a central material element that gave them flexibility to tailor their meal plan and to search for recipes for special offers before purchasing, instead of picking up an item without having a clear usage plan. The ability to tailor their meal plans enabled households to efficiently use the special offer, save money, transform such purchases from a pure to a planned impulse, and integrate them into their weekly food consumption routine.

## 5 Discussion

Acquiring food is the first stage of the consumption process for consumers and thus plays an important role in contributing to food (un)sustainability. In this study, we were interested in understanding consumers' everyday shopping routines from mainstream food retailers (in-store and online), and the ways in which material infrastructures and impulse buying behaviour had implications for household food waste. In engaging with this issue, this study shows that the challenges are deeply anchored in shopping routines, including the nature and the arrangement of consumption practice elements (i.e., meanings, competencies, and materials). The lack of shopping planning, including failing to check food stocks at home (Farr-Wharton et al., 2014), failing to make shopping lists and shopping opportunistically (Ganglbauer et al., 2013), cause shopping routines to tumble into traps of supermarket marketing cues, such as distractions of store layout design and temptations of special offers (Zhang et al., 2007; Hultén & Vanyushyn, 2011; Porpino et al., 2015). Consequently, shopping routines become increasingly marked by overprovisioning food and impulsivity, affecting households' food-waste levels (Farr-Wharton et al., 2014). Despite households' awareness of the food-waste issue, these impulses (pure and planned) were the most common causes of food waste in these households due to a lack of usage plans; foods were ultimately lost among the rhythms of a busy household. Households' time constraints are part of their dispersed practices that may affect impulse buying practices. Ganglbauer et al. (2013) discuss how shopping, as an integrated practice, is affected by dispersed practices, or as Warde (2005) describes it, exists as part of a nexus of everyday practices. Evidenced in this research are examples of how shopping practices can be disrupted by multiple dispersed practices that compete with values and circumstances, such as managing time in busy life routines (work or leisure), geographical accessibility (location), availability of storage space, and caring for the family. For example, the presence of children led households to better organise their shopping (e.g., planning, shopping online), to control their food expenditures, and to try to avoid food waste. However, caring for one's family resulted in some households' prioritising plentifulness or being a good provider of healthy food for the family, oftentimes resulting in food waste.

Organised in-store shoppers who, for example, used shopping lists, meal planning, and clear navigation strategies in-store, could, for the most part, avoid diversions from their shopping plans and thus engage in impulse purchasing to a much lesser extent than disorganised shoppers did. Where organised shoppers occasionally engaged in impulse buying, it was usually a planned impulse where they were buying meat, for example, as part of their shopping list, but because of a special offer, they purchased two packs. However, their organised household routine meant that they froze the surplus for later use; thus, their impulsivity did not result in surplus food entering the waste stream. For many organised shoppers, familiarity with the in-store layout and where products were located were essential competencies enabling an efficient shopping experience (saving shopping time) and, most importantly, avoiding browsing and distractions from marketing cues.

However, because of a focus on family togetherness and ideals of healthy food provisioning, routines were in transition, particularly for online shoppers, whose current shopping practices were in contrast to their previous shopping practices. Many of the changes in shopping practices were reconfigured from a change in the material infrastructure (i.e., the online app or website). Specifically, where routines were in transition, because of family needs or healthy eating ideals, the online infrastructure of the website/app allowed for a change in shopping practice from one that was primarily disorganised and ad hoc to one that was organised and planned, and thus met the household's new needs. Therefore, where practices are difficult to change, the elements of practice played an important role in establishing new and more sustainable practices (Fuentes et al., 2019; Rettie et al., 2012). Apart from the online infrastructure providing a convenient alternative to in-store shopping (Brashear et al., 2009), online technologies (e.g., online websites and supermarket apps) have reorganised households' shopping practices from disorganised to organised. Interestingly, this technology-enabled material element has extended consumers' agency by recruiting them to perform organised shopping practices as opposed to their usual disorganised and more impulsive shopping practices in-store, thereby impacting their households' food sustainability by reducing impulse buying and ensuring adequate stock assortment for family meals. However, due to the material infrastructure of the online platform, online shoppers, being generally disorganised in nature, may simply repeat a previous order without checking stock. This results in food surplus, particularly of healthy food. From a health sustainability perspective, having a stock of healthy food available throughout the week in the household is a positive, although ultimately, some perishable produce is thrown out before the next order arrives. Here, a lack of competence in dealing with surpluses is a barrier to increasing sustainable food consumption and avoiding food waste. Leveraging technology to ensure food is used via smart fridges and personalised meal planning apps that are integrated with smart appliances (e.g., a smart oven) may offer tailored solutions that would minimise the opportunity for food waste. Smart fridges that scan the fridge's contents, suggest recipes, and add items to a shopping list may assist in meal planning, purposeful purchasing, and ensuring that all produce can be used timely by the household.

Many of the online services, such as previous orders or favourites, and search functions, improve households' shopping competencies by saving time and enabling a more organised shopping experience by providing automated shopping lists. There is a risk of this becoming a rather sterile activity that may engender boredom and trigger variety seeking in time (or elsewhere). Nonetheless, the online shopping service may be seen as an enabler of sustainable shopping practices (e.g., organised, planned, less impulse buying, etc.). Online shopping has the potential to lessen food waste at the household level, subject to integrating all foods, including any surplus, into their weekly food routines. Practices such as checking one's food trolley/basket and using the search function to search for specific products enabled households to prevent overbuying and reduce impulse buying to a greater extent than in-store purchasing would have, for these generally disorganised food shoppers. The search function prevented households from temptations and distractions (e.g., special offers and new products), often reported in traditional supermarket shopping. This exemplifies the importance of the design of technologies and how it may rearrange and reformulate elements of shopping practice.

Notwithstanding all the above, several online marketing cues, such as viewing special offers and missed savings notifications, encouraged online shoppers to engage in impulse buying. Consistent with previous literature on e-commerce marketing, online grocery retailers may encourage impulse buying (Chan et al., 2017; Liyanage & Wijesundara, 2020). However, firstly, if households were competently including special offers into their meal plan during shopping, by searching online for recipes, they would ensure utilisation of these special offers. Overprovisioning food in this case, rather than being viewed as pure impulse buying, becomes integrated into shopping practices as part of a flexible and resourceful meal plan. This enables households not only to save money but also to expand their meal repertoires by trying new meals. Secondly, although the final basket check prior to checkout creates an opportunity for planned impulses, it also creates a space for the rational self to take a final private look (absent social risks, time, or effort usually present in an in-store setting) at the contents, to remove items that may have been added on impulse. This mechanism weakens the potential impact of impulsivity on final purchases and thus reduces waste. Crucial here, however, is the balance between uniformity and variety. Online shoppers miss the feel-good spontaneous aspect of impulse buying when shopping online (Farber, 2020). Personality characteristics, notably the impulse buying tendency and the shopping enjoyment tendency, significantly influence impulse buying in grocery retail settings (Bellini et al., 2017; Mohan et al., 2013). The question of how long these online shoppers will allow this uniformity of shopping practice to dominate over their tendencies and desires for variety and hedonistic shopping must be considered. As retailers invest more in the online shopping experience and indulge the affective desires of consumers, we may witness an increase in impulse purchasing in this space.

## 6 Conclusion

In the context of food sustainability, overprovisioning and impulse buying are notable contributors to household food waste and thus food acquisition routines play an important role in households' consumption practices. The aim of this research was to explore the shopping routines of so-called mainstream shoppers to uncover ways their shopping practices contribute to food waste. This research, through the lens of SPT, highlights the ways in which shoppers shop and how the infrastructures of the store and the online setting, and marketing cues, affect what is purchased and how it is consumed in the household.

Food provisioning infrastructures of both in-store and online settings influence related practices, such as meal planning and storing. Organised shoppers, because of their shopping plans and competencies, are less susceptible to disruptors and marketing cues in-store. If impulse purchases are made, they are usually integrated into the organised household food routines, with the use of the freezer being important to preserve any surplus. However, disorganised shoppers were more susceptible to in-store infrastructures, special offers, and other pure impulses. Where there was no clear plan to integrate these impulses into the household's food routines, food waste resulted. Interestingly, where online shoppers sacrificed variety for uniformity in their food purchases to satisfy family needs and healthy eating ideals, this sacrifice resulted in an organised (albeit facilitated) household routine and subsequently less opportunity for food waste. However, a trade-off between sustainability and healthy eating presented—repeat overprovisioning of healthy food with ambitions to improve family diets but was not routinely consumed, resulted in food waste. What does all of this mean for online food shopping, impulsivity, and food waste? Although successful in organising the weekly shopping experience and ensuring adequate food stocks and meal planning, the online infrastructure may facilitate the perpetration of healthy food being thrown out, week after week.

## References

Amos, C., Holmes, G. R., & Keneson, W. C. (2014). A meta-analysis of consumer impulse buying. *Journal of Retailing and Consumer Services, 21*(2), 86–97. https://doi.org/10.1016/j.jretconser.2013.11.004

Baker, D., Fear, J., & Denniss, R. (2009). *What a waste: An analysis of household expenditure on food*. The Australia Institute, Policy Brief No. 6. Retrieved June 16, 2021, from http://www.tai.org.au/node/1580

Beatty, S. E., & Ferrell, M. E. (1998). Impulse buying: Modeling its precursors. *Journal of Retailing, 74*(2), 169–191. https://doi.org/10.1016/S0022-4359(99)80092-X

Bellini, S., Cardinali, M. G., & Grandi, B. (2017). A structural equation model of impulse buying behaviour in grocery retailing. *Journal of Retailing and Consumer Services, 36*, 164–171. https://doi.org/10.1016/j.jretconser.2017.02.001

Berg, J., & Henriksson, M. (2020). In search of the 'good life': Understanding online grocery shopping and everyday mobility as social practices. *Journal of Transport Geography, 83*, 102633. https://doi.org/10.1016/j.jtrangeo.2020.102633

Brashear, T., Kashyap, V., Musante, M., & Donthu, N. (2009). A profile of the internet shopper: Evidence from six countries. *The Journal of Marketing Theory and Practice, 17*, 267–282. https://doi.org/10.2753/MTP1069-6679170305

Braun, V., & Clarke, V. (2006). Using thematic analysis in psychology. *Qualitative Research in Psychology, 3*, 77–101. https://doi.org/10.1191/1478088706qp063oa

Chan, T. K. H., Cheung, C. M. K., & Lee, Z. W. Y. (2017). The state of online impulse-buying research: A literature analysis. *Information and Management, 54*(2), 204–217. https://doi.org/10.1016/j.im.2016.06.001

Dash, M., & Akshaya, L. (2016). A study on the impact of visual merchandising on impulse purchase in apparel retail stores. *International Journal of Marketing and Business Communication, 5*(2), 37–44.

Denzin, N. (1970). *Sociological methods: A sourcebook*. Routledge.

Dholakia, U. M. (2000). Temptation and resistance: An integrated model of consumption impulse formation and enactment. *Psychology and Marketing, 17*(11), 955–982. https://doi.org/10.1002/1520-6793(200011)17:11<955::AID-MAR3>3.0.CO;2-J

Dobernig, K., & Schanes, K. (2019). Domestic spaces and beyond: Consumer food waste in the context of shopping and storing routines. *International Journal of Consumer Studies, 43*, 480–489. https://doi.org/10.1111/ijcs.12527

Dulsrud, A., & Jacobsen, E. (2009). In-store marketing as a mode of discipline. *Journal of Consumer Policy, 32*, 203–218. https://doi.org/10.1007/s10603-009-9104-y

Evans, D. (2012). Beyond the throwaway society: Ordinary domestic practice and a sociological approach to household food waste. *Sociology, 46*(1), 41–56. https://doi.org/10.1177/0038038511416150

Farber, E. (2020). *Online impulse buys: The newest item on the grocery industry's list*. Grocery Dive. Retrieved August 27, 2021 from https://www.grocerydive.com/spons/online-impulse-buys-the-newest-item-on-the-grocery-industrys-list/589763/

Farr-Wharton, G., Foth, M., & Choi, J. H. J. (2014). Identifying factors that promote consumer behaviours causing expired domestic food waste. *Journal of Consumer Behaviour, 13*(6), 393–402. https://doi.org/10.1002/cb.1488

Fenton-O'Creevy, M., Dibb, S., & Furnham, A. (2018). Antecedents and consequences of chronic impulsive buying: Can impulsive buying be understood as dysfunctional self-regulation? *Psychology and Marketing, 35*(3), 175–188. https://doi.org/10.1002/mar.21078

Fuentes, C. (2014). Managing green complexities: Consumers' strategies and techniques for greener shopping. *International Journal of Consumer Studies, 38*(5), 485–492. https://doi.org/10.1111/ijcs.12124

Fuentes, C., Enarsson, P., & Kristoffersson, L. (2019). Unpacking package free shopping: Alternative retailing and the reinvention of the practice of shopping. *Journal of Retailing and Consumer Services, 50*, 258–265. https://doi.org/10.1016/j.jretconser.2019.05.016

Fuentes, C., & Svingstedt, A. (2017). Mobile phones and the practice of shopping: A study of how young adults use smartphones to shop. *Journal of Retailing and Consumer Services, 38*, 137–146. https://doi.org/10.1016/j.jretconser.2017.06.002

Ganglbauer, E., Fitzpatrick, G., & Comber, R. (2013). Negotiating food waste: Using a practice lens to inform design. *ACM Transactions on Computer-Human Interaction (TOCHI), 20*(2), 1–25. https://doi.org/10.1145/2463579.2463582

Godfray, H. C. J., Beddington, J. R., Crute, I. R., Haddad, L., Lawrence, D., Muir, J. F., & Toulmin, C. (2010). Food security: The challenge of feeding 9 billion people. *Science, 327*(5967), 812–818. https://doi.org/10.1126/science.1185383

Hultén, P., & Vanyushyn, V. (2011). Impulse purchases of groceries in France and Sweden. *Journal of Consumer Marketing, 28*(5), 376–384. https://doi.org/10.1108/07363761111150026

Ilyuk, V. (2018). Like throwing a piece of me away: How online and in-store grocery purchase channels affect consumers' food waste. *Journal of Retailing and Consumer Services, 41*, 20–30. https://doi.org/10.1016/j.jretconser.2017.11.003

Iyer, G. R., Blut, M., Hong Xiao, S., & Grewal, D. (2020). Impulse buying: A meta-analytic review. *Journal of the Academy of Marketing Science, 48*, 384–404. https://doi.org/10.1007/s11747-019-00670-w

Lahath, A., Omar, N. A., Ali, M. H., Tseng, M.-L., & Yazid, Z. (2021). Exploring food waste during the COVID-19 pandemic among Malaysian consumers: The effect of social media, neuroticism, and impulse buying on food waste. *Sustainable Production and Consumption, 28*, 519–531. https://doi.org/10.1016/j.spc.2021.06.008

Liyanage, O., & Wijesundara, C. B. (2020). Online impulse buying behaviour: A review on conceptual and practice perspectives. *Global Scientific Journals, 8*(2), 3729–3765. https://doi.org/10.11216/gsj.2020.02.35479

Massara, F., Melara, R. D., & Liu, S. S. (2014). Impulse versus opportunistic purchasing during a grocery shopping experience. *Marketing Letters, 25*, 361–372. https://doi.org/10.1007/s11002-013-9255-0

Meah, A., & Watson, M. (2013). Cooking up consumer anxieties about "provenance" and "ethics." Why it sometimes matters where foods come from in domestic provisioning. *Food, Culture & Society, 16*(3), 495–512. https://doi.org/10.2752/175174413X13673466712001

Mohan, G., Sivakumaran, B., & Sharma, P. (2013). Impact of store environment on impulse buying behavior. *European Journal of Marketing, 47*(10), 1711–1732. https://doi.org/10.1108/EJM-03-2011-0110

Muruganantham, G., & Bhakat, R. S. (2013). A review of impulse buying behavior. *International Journal of Marketing Studies, 5*(3), 149. https://doi.org/10.5539/ijms.v5n3p149

Olsen, S. O., Tudoran, A. A., Honkanen, P., & Verplanken, B. (2016). Differences and similarities between impulse buying and variety seeking: A personality-based perspective. *Psychology and Marketing, 33*(1), 36–47. https://doi.org/10.1002/mar.20853

Peck, J., & Childers, T. (2006). If I touch it I have to have it: Individual and environmental influences on impulse purchasing. *Journal of Business Research, 59*, 765–769. https://doi.org/10.1016/j.jbusres.2006.01.014

Porpino, G., Parente, J., & Wansink, B. (2015). Food waste paradox: Antecedents of food disposal in low-income households. *International Journal of Consumer Studies, 39*(6), 619–629. https://doi.org/10.1111/ijcs.12207

Rettie, R., Burchell, K., & Riley, D. (2012). Normalising green behaviours: A new approach to sustainability marketing. *Journal of Marketing Management, 28*(3–4), 420–444. https://doi.org/10.1080/0267257X.2012.658840

Richter, B., & Bokelmann, W. (2017). Explorative study about the analysis of storing, purchasing and wasting food by using household diaries. *Resources, Conservation and Recycling, 125*, 181–187. https://doi.org/10.1016/j.resconrec.2017.06.006

Schmidt, K. (2016). Explaining and promoting household food waste-prevention by an environmental psychological based intervention study. *Resources, Conservation and Recycling, 111*, 53–66. https://doi.org/10.1016/j.resconrec.2016.04.006

Schneider F. (2008). Wasting food: An insistent behaviour. *Proceedings Waste: The Social Context, 8*, 1–10.

Shove, E. (2010). Social theory and climate change: Questions often, sometimes and not yet asked. *Theory, Culture & Society, 27*, 277–288. https://doi.org/10.1177/0263276410361498

Shove, E., Pantzar, M., & Watson, M. (2012). *The dynamics of social practice: Everyday life and how it changes*. Sage.

Stancu, V., Haugaard, P., & Lähteenmäki, L. (2016). Determinants of consumer food waste behaviour: Two routes to food waste. *Appetite, 96*, 7–17. https://doi.org/10.1016/j.appet.2015.08.025

Stefan, V., van Herpen, E., Tudoran, A. A., & Lähteenmäki, L. (2013). Avoiding food waste by Romanian consumers: The importance of planning and shopping routines. *Food Quality and Preference, 28*(1), 375–381. https://doi.org/10.1016/j.foodqual.2012.11.001

Stern, H. (1962). The significance of impulse buying today. *Journal of Marketing, 26*(2), 59–62. https://doi.org/10.2307/1248439

Turkyilmaz, C. A., Erdem, S., & Uslu, A. (2015). The effects of personality traits and website quality on online impulse buying. *Procedia—Social and Behavioural Sciences, 175*, 98–105. https://doi.org/10.1016/j.sbspro.2015.01.1179

United Nations. (2020). *About the sustainable development goals*. Retrieved December 17, 2020, from https://www.un.org/sustainabledevelopment/sustainable-development-goals/

Van Doorn, J. (2016). Commentary: Why do we waste so much food? A research agenda. *Journal of the Association for Consumer Research, 1*(1), 53–56. https://doi.org/10.1086/684462

Verhagen, T., & van Dolen, W. (2011). The influence of online store beliefs on consumer online impulse buying: A model and empirical application. *Information and Management, 48*(8), 320–327. https://doi.org/10.1016/j.im.2011.08.001

Verplanken, B., & Herabadi, A. (2001). Individual differences in impulse buying tendency: Feeling and no thinking. *European Journal of Personality, 15*(1 SUPPL), S71–S83. https://doi.org/10.1002/per.423

Warde, A. (2005). Consumption and theories of practice. *Journal of Consumer Culture, 5*, 131–153. https://doi.org/10.1177/1469540505053090

Warde, A. (2014). After taste: Culture, consumption and theories of practice. *Journal of Consumer Culture, 14*, 279–303. https://doi.org/10.1177/1469540514547828

Wells, J. D., Parboteeah, V., & Valachich, J. S. (2011). Online impulse buying: Understanding the interplay between consumer impulsiveness and website quality. *Journal of the Association for Information Systems, 12*(1), 32–56. https://doi.org/10.17705/1JAIS.00254

Yassin, A. C., & Soares, A. M. (2021). Buy now before it expires: A study of expiration date-based pricing. *International Journal of Retail and Distribution Management, 49*(4), 514–530. https://doi.org/10.1108/IJRDM-10-2019-0328

Zhang, X., Prybutok, V. R., & Strutton, D. (2007). Modeling influences on impulse purchasing behaviors during online marketing transactions. *Journal of Marketing Theory and Practice, 15*(1), 79–89. https://doi.org/10.2753/MTP1069-6679150106

# Making the Consumption of Food Circular: The Karma App and the Re-qualifications of Surplus Food

*Christian Fuentes, Emma Samsioe, Maria Fuentes, and Jenny Ivre*

## 1 Introduction

Digitalisation is reorganisation of contemporary food markets. New smartphone apps for the ethical consumption of food, convenient meal box services, food swapping apps, and a surge in alternative food networks enabled by digital devices and social media networks are just some of the consequences of the ongoing digitalisation of food markets.

Many of the new digital food platforms being introduced promise not only convenience or better prices, but also sustainable modes of food consumption. New digital platforms and applications, it is often argued, will enable us to consume more ecological food, shop more local food, eat

C. Fuentes (✉) • E. Samsioe • M. Fuentes • J. Ivre
Department of Service Studies, Lund University, Lund, Sweden
e-mail: christian.fuentes@ses.lu.se; emma.samsioe@ses.lu.se; maria.fuentes@ses.lu.se



A. Dulsrud, F. Forno (eds.), *Digital Food Provisioning in Times of Multiple Crises*, Consumption and Public Life,
https://doi.org/10.1007/978-3-031-46323-5_6

more plant-based, and, importantly for this chapter, reduce food waste (Southerton & Fuentes, 2022). However, to realise this potential, digital food platforms have to re-configure household practices; they have to become part of and be able to change households' shopping routines (Samsioe & Fuentes, 2022). Previous studies have shown that this is difficult as food practices often are highly routinised and held in place by socio-technical systems, norms and habits, as well as established social relations and identities (Fuentes & Samsioe, 2020).

This chapter takes a closer look at one such digital food platform—the Karma app (www.old.karma.life). The Karma app is a Swedish food platform focused on reducing food waste by allowing restaurants and retailers to sell unsold surplus food to consumers at a discount. The Karma app is one of several smartphone applications that have been launched focused on 'fighting' food waste by facilitating the selling of products with a soon-to-be-expired date (other examples available on the Swedish market are Too good to go (togoodtogo.com) and Matsmart (matsmart.se)). Consumers can download the Karma app, find surplus food, order, and pay online and then pick up the food at the restaurant and retailers. As a result, food products go from being surplus, soon to be discarded, to become valuable goods again.

What the Karma app sets out to accomplish is challenging. Karma both proposes a new mode of food acquisition and a new way to value food. Karma asks of consumers not only to shop food in a new way but also that consumers change the way they value a central quality of food—freshness. It does so by re-qualifying food-soon-to-be-expired as edible, sustainable, and economical (see also, Petersson Ivre, 2020). By doing this, Karma enables the circular consumption of food, i.e., it enables and encourages consumer activities that prevent food waste and support actors involved in circular business and supply chains (Heidenstrøm & Hebrok, 2022; Massimiliano & Luigi, 2021; O'Neill et al., 2022).

Previous research has shown that efforts to change the way we shop for food often fail due to problems associated with the app/digital platforms or because the new practice promoted does not manage to replace existing and already routinised shopping modes (Fuentes et al., 2021). Furthermore, norms of what is 'proper' can often stand in the way of food waste reduction (Evans, 2011, 2012) and be difficult to change (e.g., Halkier, 2017; Jackson et al., 2018).

To try and make sense of Karma's efforts to promote circular food consumption, we will combine the resources of constructivist market studies

(CMS)—more specifically the concepts of market device and qualification—and the sociology of consumption (for a similar approach, see Fuentes & Samsioe, 2020; Petersson Ivre, 2020). Drawing on these concepts we ask: How and under what condition is the re-qualification of food soon-to-be-expired possible? And in what way, if any, is this circular food consumption also an example of sustainable food consumption?

The analysis is mainly based on digital observations of the Karma apps and its online marketing and interviews and digital walkthroughs with ten households using the Karma app.

Drawing on this material and the concepts of market device and qualification we show that while the Karma app does, under certain circumstances, accomplish the task of promoting and enabling a new mode of food shopping and a new way of framing food, this is a messy process with uncertain results. While Karma works tirelessly, through its app design and marketing, to qualify soon-to-be-expired food as a resource for sustainable food practices, it has little control over the multiple re-qualifications performed by consumers. In the household practices of consumers, the food sold via the Karma app becomes re-qualified. Here sustainability aspects of this mode of food consumption commonly fall in the background as issues of enjoyment, novelty or thrift are highlighted. The chapter ends by discussing some of the sustainability and social implication of this process of re-qualifications. Critically, the analysis shows that while the mode of food consumption accomplished can be considered circular, it is not necessarily sustainable.

## 2 Digital Platforms, Re-qualifications, and Food Consumption

As previously stated, we here combine the theoretical resources from constructivist market studies with insights from the sociology of consumption to conceptualise how the Karma app as a market device works to re-qualify soon-to-expire food as edible, sustainable, and economic and how this food is then re-qualified by the consumers making use of the app.

We begin, however, with a brief introduction of these fields and an argument for their combination. The field of constructivist market studies is an interdisciplinary field of research heavily influenced by economic sociology that takes an interest in the practical and material organisation of markets (Kjellberg & Helgesson, 2007). Within constructivist market studies, neither markets (in the plural) nor their mechanisms are taken as given. Through empirical studies scholars are interested in tracing the practices and processes by which markets are socio-materially organised. Studies within this field have, for example, explored the role of economic theories in performing markets (Mason et al., 2015), the configuration of markets actors (Andersson et al., 2008), how market devices of different sorts work to shape markets (Callon et al., 2007), and the qualification and valuation of products (Dubuisson-Quellier, 2007). However, while the issue of consumption is at times mentioned in this field, few studies attempt to connect markets and their devices with the field of consumption.

Conversely, the sociology of consumption makes consumption its focal point of attention (as the name so aptly communicates). This interdisciplinary field emerged in the 80s as a response to a perceived neglect of consumption and an overemphasis on production (Miller, 1995). While originally focused on the role of advertisement and marketing in the production of consumer culture (see, e.g., Falk & Campbell, 1997; Featherstone, 1991; Miller et al., 1998), sociology of consumption is today much more interested in household consumption and the consumption of environmentally problematic 'resources' such as energy (Evans, 2019). This shift is in part due to the practice turn in consumption. As theories of practice have made their way into the field, they have come to redefine the field, shifting focus away from more conspicuous and spectacular forms of consumption and onto more mundane practices; away from the purely cultural and towards practical and material aspects of consumption (Shove et al., 2012; Warde, 2005, 2014). While this shift has been important for the development of the field and allowed scholars to move beyond 'culturalist' approaches to consumption, it has also meant that the role of 'commercial actors' in the shaping of consumption is now seldom studied (Evans et al., 2017). Thus, while scholars working in the field of constructivist market studies often neglect to study consumption in their efforts to understand the socio-material organisation of markets, scholars within the field of sociology of consumption tend to ignore the role that commercial actors, logics, and devices play in shaping consumption (Dubuisson-Quellier, 2022; Fuentes & Samsioe, 2020).

The theoretical approach developed here is thus positioned in between these two fields. It is an effort to bridge these two fields and bring the study of markets and consumption together. In what follows, we explore how the Karma app as a market device shapes the consumption of soon-to-be-expired food products by re-qualifying them. However, to understand this, we must also understand how food products qualified by Karma are re-qualified by consumers as they are shopped, cooked/prepared, and eaten as part of the households' interconnected food practices.

Two concepts are thus of particular importance here: that of market device and that of qualification. Market device is a concept used in the field of constructivist market studies to denote socio-material arrangements that take part in the making and shaping of markets (Callon et al., 2002). Market devices are material-semiotic arrangements that mediate and afford consumers' understandings and valuation of goods. Examples are packaging (Cochoy, 2004) or supermarket displays (Cochoy, 2010) that shape how consumers understand and value a good. Market devices do this through 'qualification', i.e., through continuous work that ascribe goods with a set of qualities that make them comparable to other goods. By qualification we are referring to the process by which an entity—in this case food—acquires its qualities. The underlying notion of this concept is that the qualities of a good are not given, not 'natural' or innate, but instead the result of an active and continuous processes of qualification.

In this case, the concept is used to put emphasis on the ways in which the Karma app actively works to shape the way food is framed, shopped, and consumed. By seeing the Karma app as a market device, we can shed some light on its economic performativity.

As this also indicates, this chapter takes a distinct socio-material approach. The Karma app—as a market device—is a generative practice element. In this analysis, devices are actively involved in the reproduction and shaping of practice (Preda, 1999). An app is therefore in this framing not necessarily a passive tool but rather to be understood as a potentially active co-performer of practices (Fuentes et al., 2021). We will, in the following analysis, make use of the concept of qualification to clarify how a market device—such as the Karma app—shapes both practices and practice resources as food and how consumers then re-qualify both the app and the food it provides.

The qualification process is often complex. It is not necessarily linear nor has it a clear end point. To take a common example, the development of consumer products often involves various qualification processes (Dubuisson-Quellier, 2010). Products are developed, tested in labs, field tested, redefined by advertisement agencies, qualified in retailing practices.

Qualification, however, does not stop there. Products are typically also re-qualified by consumers when they talk about the products, use them in everyday practices, and eventually also dispose of them (Fuentes & Fuentes, 2017).

As this also illustrates, the 'work' of qualification is performed by multiple entities and actors and is part of numerous practices. To be specific, qualification involves the material-semiotic transformation of goods. Qualifying products commonly involves both altering their materiality—e.g., through production and design—and the meanings attached to them—e.g., through advertisement.

Furthermore, qualifications are commonly contested. This is important for our subsequent analysis. The actors involved in the qualification process do not always share the understanding of a product's qualities, and products are also commonly re-qualified as they become part of various practices (Fuentes & Fuentes, 2017). This does not mean that the qualification process is broken or interrupted. As Dubuisson-Quellier (2010) makes clear, there is no need for a 'consensual agreement on a product's qualities for exchange to take place' (p. 75). The same products can therefore be qualified differently by different actors and devices across time and space.

Combining these concepts, we will explore below how Karma—as a market device—re-qualifies soon-to-be-expired food as to make it valuable again by not only promoting a new mode of food shopping but also framing the consumption of this food as something that is simultaneously sustainable, economical, and fun. We will furthermore delve into what happens as the Karma app and the food it (re)commercialises becomes part of and links to consumers' food and non-food related practices.

## 3 A Study of Karma

In this chapter, we will draw on material collected on the Karma app. The material analysed here is part of the broader study of digital food platforms and the promotion of sustainable consumption done by the Swedish team as part of the PLATEFORMS project.

In our study, Karma was one of three sampled cases (meal box schemes and Reko-rings were the other two). Karma was chosen because it represented an example of the phenomenon under study—digital platforms designed and developed to re-configure the relationship between production and consumption in a sustainable direction (Southerton & Fuentes, 2022).

As we have explained in previous publications, Karma is a food aggregator platform that aims to market soon-to-expire food (Samsioe & Fuentes, 2022). It was developed by a Swedish start-up in 2016 and expanded to the UK in 2018 and France in 2019 (https://karma.life/about). Karma had at the time of the study more than 500 000 users and 2000 food sellers connected in several cities across Europe.

The goal of this platform is to sell 'surplus food' from coffee shops, restaurants, and retailers to consumers at a reduced price. By doing this, consumers can 'rescue' food and these outlets can receive revenue for what would otherwise have been waste. The aim, Karma argues, is to promote sustainability by reducing food waste.

To understand how the Karma app works to qualify food through its design and marketing, we carried out a series of digital observations of the app itself. The objective was to 'trace' the Karma apps' multiple functions and the messages it conveyed on and through the platform. We also conducted observations of Karma's webpage and its social media accounts (domestic and international) on Facebook. The digital observations were conducted between 1 March and 28 October 2019 and were documented via means of screenshots. Informed by the socio-material approach outline above, we read the 'script' of these digital entities, interested both in what actions they enabled and how these actions were framed as meaningful to users (Fuentes & Sörum, 2019).

To understand how the Karma app was used by consumers, we also conducted ethnographic interviews with ten households that used the Karma app (see Table 1). We were interested in households that had

**Table 1** Research participants

| *Pseudonym* | *Household type* | *Age interval* | *Gender* |
|---|---|---|---|
| GO | Couple | 40–49 | Female |
| MA | Single | 40–49 | Male |
| AN | Family | 40–49 | Female |
| SO | Family | 50–59 | Female |
| MJ | Couple | 30–39 | Male |
| BE | Family | 30–39 | Female |
| PA | Family | 50–59 | Female |
| JN | Family | 40–49 | Male |
| AD | Couple | 30–39 | Male |
| AG | Single | 20–29 | Female |

managed to integrate the Karma app and use it to, more or less, habitually purchase food. The interviews were conducted over a period of 16 months from June 2019 to October 2020. All the interviews except one were conducted digitally, due to the Covid-19 pandemic.

The interviews covered both the households' everyday food practices broadly—the shopping, cooking, eating, and disposing of food—and the specific use of the Karma app. During the interviews the participants were also asked to show the interviewer the Karma app and demonstrate its use. The interviewer and the participants would go through the app and talk about its multiple functions and how and to what extent these were used by the participant. The interviews were audio and video recorded and transcribed in full.

The screenshots resulting from the digital observation and the transcription from the interview were analysed using the qualitative analysis software Nvivo. Central analytical questions were the following: How does Karma promote its specific mode of food shopping? How is the food sold qualified by the Karma app and marketing material? How and why is the Karma app used by households? How is the food bought through Karma re-qualified by consumers? In what follows we present the results of the analysis using interview quotes as well as summaries of the observations conducted.

## 4 Karma in Everyday Life: Unfolding Circular Food Practices

As mentioned previously, Karma is an aggregator app that interconnects multiple coffee shops, restaurants, and retailers, enabling these outlets to sell their soon-to-be-expired food at a discount through the platform. By doing this, the app works to reduce food waste by enabling a circular mode of food consumption.

Below we will first look closely at this market device and describe how it works to make both possible and meaningful a new form of circular food shopping. This is followed by an analysis of how the Karma app is used by the households that participated in our study.

### *Understanding the Karma App and Its Qualification of Surplus Food*

The Karma app is designed to enable consumers to purchase soon-to-be-expired food from multiple outlets. Inscribed into its design are easy step-by-step instructions for consumers on how and why to use the Karma app. The app encourages consumers to (1) download and install the app, (2) start browsing soon-to-be-discarded food, (3) 'rescue' said food at half the price, (4) pick up the food also using the app, and, finally, (5) 'enjoy' (www.karma.life/ 13 June 2019).

As consumers log into the app, they are presented with the start menu. The start menu encourages consumers to explore the various food outlets connected to Karma available in their area. It is possible for consumers to modify the map and browse this interactive app based on preference. They can choose to have a meal, a coffee, or shop for groceries, all through Karma. It is also possible to tap on any of the restaurants, cafés, or grocery stores visible on the map. Tapping them generates information regarding the food venue, the available food at the specific time, and for how long the food item is available (before being discarded). Consumers are additionally given the opportunity to configure the map by choosing to show only vegetarian, vegan, lactose-free, or gluten-free meals and food. Furthermore, consumer can choose to follow specific food venues—a favourite restaurant, for example—and receive notifications when special offers or specific types of food are available. When consumers find a meal or other food item to purchase, the app facilitates this quickly and smoothly. Consumers pay through the app and receive a confirmation and a time slot to pick up the item. When the consumer goes to pick up the item, she or he shows the staff the Karma app and the confirmation page to receive the products they purchased. On completing the transaction, consumers are given positive feedback as the app generates a 'thank you for rescuing food' message before asking them for feed-back, promoting them to rate the purchase.

As this short description of the functions of the Karma app indicates, this app is mainly designed to enable and encourage a new mode of food shopping (Fuentes et al., 2021). It works to accomplish this through both its specific physical design, which enables certain user actions, and through the texts and imagery on the app which frames those actions as meaningful in different ways. The Karma app is clearly designed to enable Karma shopping: a specific mode of food provisioning. The apps design works towards recruiting consumers and involving them in this new mode of food shopping.

However, in the process of trying to enact this new mode of food provisioning, the Karma app also works towards re-qualifying food. It works to transform soon-to-be-discarded food. To go from less valuable, on the verge of inedible foodstuff, to edible, enjoyable, economic, and sustainable food. This process of re-qualification involves linking the Karma food to various, supposedly positive, qualities. Most visible, on the app and in the marketing material, is the connection to sustainability. Karma frames the consumption of soon-to-be-discarded food as a form of sustainable consumption.

To accomplish this, Karma connects to the sustainability problem associated with food waste. This is a problem that has received considerable attention, both in policy and in scholarly discussions (Evans, 2012; Hebrok & Heidenstrøm, 2019). Consumers who visit the Karma webpage can read: 'Every year, 1.3 million tonnes of food are thrown away in Sweden. The food thrown away corresponds to emissions of 2 million tonnes of CO2. About half of this is food waste—good food that could have been eaten' (www.karma.life/ 13th of June 2019).

Karma draws attention to the important problem of food waste only to immediately offer a convenient solution to this problem. The Karma platform, together with the food venues linked to the platform and the consumers using the platform, are positioned as 'food rescuers', working to reduce food waste and contributing to sustainability.

This message runs through the design of the app. The first thing a user sees when opening the app is the text 'a million rescued food items!' with the Karma logo in the background. Consumers are also urged to join the 'food rescuing gang!' and/or invite friends to join the community of food rescuers. Consumers are furthermore encouraged to read their Karma profile where they can find information regarding their order and the food they have 'rescued'. The app keeps track of quantifies and visualises the users' sustainability impact. In the design and marketing of Karma, food waste is a problem that can be easily addressed; just buy more Karma food! Karma also links itself to circular consumption, campaigns to reduce food waste, and numerous sustainability events, thereby contributing to the qualification of Karma food as sustainable food.

However, this is not the only quality that is attached to the food. Karma food is also framed as economical. In the design and marketing of Karma, there is no contradiction between inexpensive and sustainable food. On the contrary, Karma phrased the buying of soon-to-be-expired food at discount as a typical win-win circular consumption scenario (Ariztia & Araneda, 2022). The Karma slogan: 'Buy good food at half price! Stop good food from being thrown away!' illustrates this app's core message. Karma, through its design and marketing work, offers consumers the opportunity to save money and combat food waste at the same time (Fuentes et al., 2021).

Finally, Karma also works to qualify its food as edible, tasty, and enjoyable. It counteracts the preoccupation with freshness and question consumers' understanding and practices surrounding best-before-dates. On Karma's Facebook page one can read:

> Best-before-dates includes a time margin to the day when the product does not maintain the same quality as when it was packaged. So, instead of throwing it away based on its best-before-date, smell and taste [the food]! Share this post to remind your loved ones. (www.facebook/Karma/ 27 October 2019)

Karma thereby 'educates' consumers on the fact that best-before-date does not necessarily mean that the food has gone bad after the stipulated date. Karma also uses aesthetically pleasing images to illustrate the Karma food on offer. Beautiful images of sushi or cinnamon buns are used to qualify Karma food as pleasurable food. Texts that describe the food as 'yummy' or 'enjoyable' are also common on the Karma's app, webpage, and Facebook accounts.

In summary, Karma works, through its app design and online marketing, to enable and encourage a new mode of food acquisition and thereby also supposedly a more circular mode of food consumption. To accomplish this, it does not only enable a new shopping practice, but it also works to (re)qualify the food it sells. Countering the widespread notion that 'old' food or food close its best-before date is inedible, stale, and therefore to be avoided, Karma qualifies its food as sustainable, economical, and edible/enjoyable (see, also, Fuentes et al., 2021; Petersson Ivre, 2020).

### *Consumers and the Re-qualifications of Karma Food*

How is the Karma app used? And how is Karma food valued by consumers? It is, as previous studies within Science and Technology Studies (STS), practice theory, and consumption more generally have shown, both possible and probable that users deviate from the scripts of artefacts (e.g., Fuentes & Sörum, 2019; Hand & Shove, 2007). This is also what we observe in our material. Examining the food practices of households that participated in our study, one can note that there were multiple ways in which consumers re-qualified Karma food.

*Karma Food Consumption as an Economic Practice: The Hunt for Inexpensive Food*

In some situations, the use of the app and the consumption of Karma food is mainly re-qualified as an economic practice. For some, the Karma app provides them with access to the food they normally consume but at a discount. For example, if they typically buy lunch from a specific restaurant, they make a habit to check the Karma app before ordering to see if they can find a meal from that restaurant at a discount: 'One might also become a bit stingy and start to think that it is unnecessary to go and buy a full price meal when you can buy a perfectly fine meal from yesterday for half the price' (AN 40–49, female, Karma consumer).

In other instances, the discounted price may make food available to households that would otherwise refrain from consuming because of its price. However, it is important to point out that in our sample it was seldom the case that consumers could not afford the item in absolute terms. Rather they choose to refrain from a specific type of food consumption which is relationally more expensive, guided by notions of thrift. In these cases, Karma, and the discounted price, offered consumers the opportunity to buy a 'treat' while still consuming in line with their ideals of thriftiness:

> It has mainly been bread. At the Gateau (bakery) down by the station. And the one in the sales hall. I have been to the one at the station many times. Instead of paying 60 crowns for a piece of bread, I can pick it up for 30 in the afternoon. Because it is a more exclusive type of bread one could say. / Yes, and then it could be a bit of a treat you know. Mabey you are expecting visitors and then I think "yes I will go pick up some fika, some nice bread". (GO 40–49, female, Karma consumer)

Buying Karma food at a discount can also be framed as way of being economically responsible, or a way to avoid 'a guilty conscious' brought on by perceived overspending.

> And maybe I have dined out a bit too often lately. Well, maybe then I can appease my guilty conscience by using this app. In other words [I use the app] to appease my guilty conscience when I am out of money. I eat out a lot. (AG 20–29, female, Karma consumer)

As these examples illustrate, in our sample, consisting mainly of well-off middle-class households, being economical is seldom a necessity but rather a choice. In this context, Karma food is mainly a bargain, or a treat and Karma consumption is mainly an economic practice.

*Karma Food Consumption as an Enjoyable Practice: Fun, Enjoyable, and Novel Food Shopping*

Karma food consumption is also commonly qualified by consumers as a fun, enjoyable, and novel practice. Because of the connection to various food venues, Karma food is at times qualified as a channel for enjoyable experiences. Angela, for example, mainly associated Karma with Espresso House and a couple of its products she regularly bought using the Karma app:

> Espresso House had some decent sandwiches. Grilled avocado, these sandwiches were delicious. And then I bought a mini bowl. Yes, something like that. I think it was a lot of Espresso House. But I remember also thinking that I could buy fika there. (AG 20–29, female, Karma consumer)

However, she also values the app because it offers a convenient and fast way to shop for food:

> Yes. So, I mostly thought that the app, that I can use the app spontaneously. Not much planning. It is not like I am planning to use the Karma app tomorrow. For me, it should be now I am hungry, so I will use the app. It should be fast. (AG 20–29, female, Karma consumer)

Here, rather than being an unusual and complicated way of shopping for food, Karma becomes the convenient channel. As a food aggregator app, it allows consumers to browse and order from various food venues on the same app and using a helpful map.

Other consumers bring the novel quality of Karma to the fore when discussing the app and the food they purchase. Consumers, and particularly those with an interest in technology, find it exciting to try out a new app and practice. They become enthusiastic over the app and will often talk about the app with people in their social network, promoting it: 'It's fun. Maybe I should point out that I like all [apps], so it's pure curiosity for me. Well, the savings—whatever. To be honest, I don't find the sustainability argument convincing … So, it's a fun thing to use' (MA 40–49, male, Karma consumer).

The novelty aspect typically also leads them to make more purchases in the beginning. Excited about the app they stay informed, checking notification and exploring different restaurant and stores in Karma's interactive map. Over time, however, the novelty aspect wears off, and consumers use of the app and purchases made tend to become less frequent.

*Karma Food Consumption as a Sustainable Practice: Rescuing Food to Save the Environment*

Finally, Karma food shopping is also at times qualified as a sustainable mode of food shopping. There were multiple examples in our material of Karma food shopping and Karma food being qualified as sustainable by consumers. For some of our participants, Karma food consumption is primarily a sustainable mode of consumption. In these cases, Karma is often one of the multiple ways in which consumers attempt to practice a sustainable everyday life. This can include other sustainability practices such as package-free shopping, buying second-hand clothes, or riding the bike rather than taking the car. Karma food consumption is thus for these consumers part of a set of sustainability practices they perform:

> But sometimes when I don't know what we will have I check [the app] anyway. Most often at the beginning of the week. And then it can happen during the weekend as well. And then I take the bike there pick up this food. And then I feel good about myself. Thrifty, like a save the planet-person. (PA 50–59, female, Karma consumer)

In the example above we see how Petra combines at least two sustainable practices (avoiding the car and shopping via Karma) in an effort to enact a more sustainable lifestyle. For these consumers, Karma fits into a broader ambition; to reconfigure their everyday practices in a more sustainable direction.

Karma can also be specifically connected a long-standing commitment to reducing food waste:

MN: Nothing goes to waste in our household. It just doesn't happen. So you also want to contribute to reduce waste in the rest of the world.

Interviewer: OK but why is this so important for you? What makes it important?

MN: I grew up with the whole concept. You eat everything on your plate. Nothing is to go to waste. I am not the only one who grew up this [ideal]. I think that it is something imprinted in many of us, and in me in particular. No, I am unable to throw out even the smallest piece [of food]. (MN 30–39, male, Karma consumer)

In the interview extract above, Martin explains that avoiding food waste is a practice that he learned early on in his life. Karma is thus merely an extension of this practice. Similarly, another of our participants works with waste reduction and sees her use of Karma as an extension of her mission to reduce waste in different areas. Her professional mission becomes also a personal one as she tries to align her private consumption with her work practices.

In these cases, Karma fits into a long-standing life project that typically involves not just buying from Karma but also being mindful of the food you buy in general, not throwing away food just because it is past its best-before date, and planning your cooking as not to generate food waste. Here the Karma app and Karma food are resources in the enactment of a sustainable lifestyle.

### *Karma and Circular Food Consumption*

What does this mean for the Karma app's ambition to reduce food waste by rescuing food? Does the Karma accomplish its mission to make food consumption circular?

The study shows that Karma does indeed make food consumption circular in the sense that it 'recirculates' food that would otherwise have perished. The Karma app, as we showed, successfully enables a new mode of food consumption in which soon-to-expire food is sold at a discount, thereby 'rescuing' it from becoming food waste. The 'buffe' of qualities offered by Karma makes this mode of food consumption meaningful for consumers in various ways.

The way in which Karma consumption was performed by the participants in our study was not unproblematic, however. While the Karma food platform was in some cases used as a resource in the enactment of sustainable lifestyles, it was in many other cases valued not primarily for its sustainability benefits but rather for its ability to offer discounts on food, because of its novelty aspects, or the convenience it offered.

Many would argue that this is unproblematic. Indeed, it is not uncommon for scholars and advocates of sustainable consumption to argue that the most feasible way to promote a shift towards less resource-intensive practices is by foregrounded other, non-sustainability qualities (e.g., Heidenstrøm & Hebrok, 2022; Meyer, 2001). While this pragmatic argument has its merits, it tends to ignore the importance of performativity. When Karma food shopping is re-qualified as mainly an economic, convenient, enjoyable, or novel practice, the Karma app and Karma food become instead enacted as resources for bargain hunting, treat consumption, convenient food consumption, or novelty shopping. So, while all Karma shopping contributes to the rescuing of food, these latter performances of this practice do little to enact sustainability. They do little to raise awareness or interest in food waste reduction or to inspire or promote sustainability change in other practices.

Taking the argument one step further, one can claim that not only is sustainability excluded from these performances. When Karma shopping becomes only or primarily about price, fun, or convenience, consumers are enacting modes of food consumption that often counteract the transition to sustainability. When food is mainly bought to make a bargain, because it is convenient, or in search of novelty aspects, there is a risk of increasing levels of food consumption or at the very least locking-in already high levels of consumption (Petersson Ivre, 2020). This, in turn, also runs the risk of normalising the overproduction of food. If the surplus can be sold via a new channel, what motivation is there for food venues to better plan their food production and food retail?

## 5 Conclusions

In this chapter we empirically examined and conceptualised how Karma—a digital food platform—enabled the 'rescuing' of food waste by re-qualifying food close to its expiration date as sustainable, convenient, and enjoyable food. We showed that while Karma qualified its app and food as sustainable *and* economical, convenient, and enjoyable, consumers commonly re-qualified Karma consumption as *primarily* an economical, convenient, or enjoyable mode of food consumption, thereby downplaying or even excluding the quality of sustainability. This, we argued, is problematic as Karma food consumption thus becomes principally an economic, convenient, or hedonistic mode of shopping, foregoing its sustainability potential and possibly even leading to the overconsumption of food.

This analysis makes three contributions. First it reminds us of the importance of being sceptical of sustainability claims made by various (commercial) actors. If, how, and to what extent a mode of consumption or a specific digital food platform is sustainable is seldom clear cut. What may seem like an example of a more sustainable mode of consumption can have complicated and ambiguous effects when examined more closely (see also, Cajic et al., 2022).

Second, this analysis shows us the importance of examining the entanglements of digital food platforms and consumption practices. Without a contextualised understanding of how digital food platforms are integrated into households' everyday (food) practices, we can neither develop an informed analysis of the platforms 'successful' integration, nor can we fully evaluate their performative (sustainability) effects (for a similar argument, see, Fuentes & Samsioe, 2020).

Finally, the chapter also sets out to develop and illustrate the usefulness of combining concepts from economic sociology and sociology of consumption to understand the digital shaping of food consumption. By making use of the concepts of market device and qualification, we were able to take into account the socio-material 'work' performed by this digital food platform to qualify Karma shopping as a specific type of practice and Karma food as a specific type of food. Furthermore, by also following the qualification process into consumption, we were able to show the discrepancy between qualifications and discuss what this meant in this specific case.

## References


Andersson, P., Aspenberg, K., & Kjellberg, H. (2008). The configuration of actors in market practice. *Marketing Theory, 8*(1), 67–90.

Ariztia, T., & Araneda, F. (2022). A 'win-win formula': Environment and profit in circular economy narratives of value. *Consumption Markets & Culture, 26*(1), 1–23.

Cajic, S., Brückner, M., & Brettin, S. (2022). A recipe for localization? Digital and analogue elements in food provisioning in Berlin—A critical examination of potentials and challenges from a gender perspective. *Sustainable Production and Consumption, 29*, 820–830.

Callon, M., Méadel, C., & Rabeharisoa, V. (2002). The economy of qualities. *Economy and Society, 31*(3), 194–217.

Callon, M., Millo, Y., & Muniesa, F. (Eds.). (2007). *Market devices*. Blackwell.

Cochoy, F. (2004). Is the modern consumer a Buridan's donkey? Product packaging and consumer choice. In K. M. Ekström & H. Brembeck (Eds.), *Elusive consumption*. Berg.

Cochoy, F. (2010). 'How to build displays that sell'. The politics of performativity in American grocery stores (*Progressive Grocer*, 1929–1946). *Journal of Cultural Economy, 3*(2), 299–315.

Dubuisson-Quellier, S. (2007). The shop as market space. The commercial qualities of retail architecture. In D. Vernet & L. de Wit (Eds.), *Boutiques and other retail spaces* (pp. 16–45). Routledge.

Dubuisson-Quellier, S. (2010). Product tastes, consumer tastes: The plurality of qualification in product development and marketing activities. In L. Araujo, J. Finch, & H. Kjellberg (Eds.), *Reconnecting marketing to markets* (pp. 74–93). Oxford University Press.

Dubuisson-Quellier, S. (2022). How does affluent consumption come to consumers? A research agenda for exploring the foundations and lock-ins of affluent consumption. *Consumption and Society, 1*(1), 31–50.

Evans, D. (2011). Blaming the consumer—Once again: The social and material contexts of everyday food waste practices in some English households. *Critical Public Health, 21*(4), 429–440.

Evans, D. (2012). Beyond the throwaway society: Ordinary domestic practice and a sociological approach to household food waste. *Sociology, 41*(1), 41–56.

Evans, D. (2019). What is consumption, where has it been going, and does it still matter? *The Sociological Review, 67*(3), 499–517.

Evans, D., Welch, D., & Swaffield, J. (2017). Constructing and mobilizing 'the consumer': Responsibility, consumption and the politics of sustainability. *Environment and Planning A, 49*(6), 1396–1412.

Falk, P., & Campbell, C. (Eds.). (1997). *The shopping experience*. Sage.

Featherstone, M. (1991). *Consumer culture and postmodernism*. Sage Publications.

Fuentes, C., & Fuentes, M. (2017). Making a market for alternatives: Marketing devices and the qualification of a vegan milk substitute. *Journal of Marketing Management, 33*(7–8), 529–555.

Fuentes, C., & Samsioe, E. (2020). Devising food consumption: Complex households and the socio-material work of meal box schemes. *Consumption Markets & Culture, 24*(5), 492–511.

Fuentes, C., & Sörum, N. (2019). Agencing ethical consumers: Smartphone apps and the socio-material reconfiguration of everyday life. *Consumption Markets & Culture, 22*(2), 131–156.

Fuentes, C., Cegrell, O., & Vesterinen, J. (2021). Digitally enabling sustainable food shopping: App glitches, practice conflicts, and digital failure. *Journal of Retailing and Consumer Services, 61*, 1–9.

Halkier, B. (2017). Normalising convenience food? *Food, Culture & Society, 20*(1), 133–151.

Hand, M., & Shove, E. (2007). Condensing practices: Ways of living with a freezer. *Journal of Consumer Culture, 7*(1), 79–104.

Hebrok, M., & Heidenstrøm, N. (2019). Contextualising food waste prevention—Decisive moments within everyday practices. *Journal of Cleaner Production, 210*, 1435–1448.

Heidenstrøm, N., & Hebrok, M. (2022). Towards realizing the sustainability potential within digital food provisioning platforms: The case of meal box schemes and online grocery shopping in Norway. *Sustainable Production and Consumption, 29*, 831–850. https://doi.org/10.1016/j.spc.2021.06.030

Jackson, P., Brembeck, H., Everts, J., Fuentes, M., Halkier, B., Hertz, F. D., et al. (2018). *Reframing conveneince food*. Palgrave Macmillan.

Kjellberg, H., & Helgesson, C.-F. (2007). On the nature of markets and their practices. *Marketing Theory, 7*(2), 137–162.

Mason, K., Kjellberg, H., & Hagberg, J. (2015). Exploring the performativity of marketing: Theories, practices and devices. *Journal of Marketing Management, 31*(1–2), 1–15.

Massimiliano, B., & Luigi, C. (2021). Transitioning into circular food consumption practices: An analytical framework. In A. Stefanakis & I. Nikolaou (Eds.), *Circular economy and sustainability* (pp. 385–407). Elsevier.

Meyer, A. (2001). What's in it for the customers? Successfully marketing green clothes. *Business Strategy and the Environment, 10*, 317–330.

Miller, D. (Ed.). (1995). *Acknowledging consumption*. Routledge.

Miller, D., Jackson, P., Thrift, N., Holbrook, B., & Rowlands, M. (1998). *Shopping, place and identity*. Routledge.

O'Neill, C., Hashem, S., Moran, C., & McCarthy, M. (2022). Thou shalt not waste: Unpacking consumption of local food. *Sustainable Production and Consumption, 29*, 851–861.

Petersson Ivre, J. (2020). *No good deed goes un-scripted—An ethnographic study on the sustainable smartphone application Karma*. (Master). Lund Univeristy, Lund.

Preda, A. (1999). The turn to things: Arguments for a sociological theory of things. *The Sociological Quarterly, 40*(2), 347–366.

Samsioe, E., & Fuentes, C. (2022). Digitalizing shopping routines: Re-organizing household practices to enable sustainable food provisioning. *Sustainable Production and Consumption, 29*, 807–819.

Shove, E., Pantzar, M., & Watson, M. (2012). *The dynamics of social practice: Everyday life and how it changes*. Sage.

Southerton, D., & Fuentes, C. (2022). Digital platforms and sustainable food consumption transitions. *Sustainable Production and Consumption, 29*, 805–806.

Warde, A. (2005). Consumption and theories of practice. *Journal of Consumer Culture, 5*(2), 131–153.

Warde, A. (2014). After taste: Culture, consumption and theories of practice. *Journal of Consumer Culture, 14*(3), 279–303.

# From Grassroots to Platforms: How Digitalisation Reconfigures Learning and Engagement with Food

*Francesca Forno, Suse Brettin, Christopher Moran, and Sandra Cajic*

## 1 Introduction

Throughout the past century, industrialised food systems have ensured mass access to fairly priced and varied food for many people, especially the middle and working classes of the Global North. However, this expansive power has proven, over time, to be unsustainable (Moore, 2015) in both

F. Forno (✉)
Department of Sociology and Social Research, University of Trento, Trento, Italy
e-mail: francesca.forno@unitn.it

S. Brettin • S. Cajic
Thaer-Institute of Agricultural and Horticultural Sciences, Humboldt-University Berlin, Berlin, Germany
e-mail: suse.brettin@gender.hu-berlin.de; sandra.cajic@hu-berlin.de





A. Dulsrud, F. Forno (eds.), *Digital Food Provisioning in Times of Multiple Crises*, Consumption and Public Life,
https://doi.org/10.1007/978-3-031-46323-5_7

ecological and socio-economic terms, involving the uprooting of traditional forms of agriculture and food sovereignty, the hyper-exploitation of human labour, the massive employment of ecologically damaging farming techniques, and threats to human health (Marsden & Sonnino, 2012).

In response to these challenges, calls for a transition to a more ecological and resilient food system based on socio-economic and gender equity, access, security, health, and well-being have begun to garner support across broad sectors of society. This in turn has given rise to important social retro-innovations, intended as 'developing knowledge and expertise that combines elements and practices from the past and the present and configures these elements for new and future purposes' (Stuiver, 2006, p. 150). In this regard, community-organised forms of procurement, generally labelled 'alternative food networks' (AFNs), have been central to this effort. AFNs result from mobilisations of civil society to counter the hegemony of the modern food regime. Their main modus operandi consists of shortening supply chains, promoting ethical small-scale farming and quality seasonal foods at fair prices, and building meaningful relationships of mutual support and trust between producers and consumers (Goodman et al., 2012; Thorsøe & Kjeldsen, 2016).

AFNs such as farmers markets, box schemes, solidarity purchasing groups, community-supported agriculture (CSA), food assembly hubs, and food-sharing initiatives have established themselves as important spaces of innovation and apprenticeship for the promotion of more environmentally, socially, and economically 'sustainable' forms of production and consumption (Dwiartama & Piatti, 2016; Miralles et al., 2017; Zoll et al., 2018). These networks have been proven to help consumers gain a better understanding of food and food systems by facilitating 'learning by doing'. As research on solidarity purchasing groups, for example, has shown, within these networks members not only find information and mutual support but also learn, build relational ties, and gain an enhanced potential to act (Forno et al., 2015). In other words, by bringing together producers and consumers in spaces that enable direct interaction, AFNs

C. Moran
Department of Management and Marketing, University College Cork, Cork, Ireland
e-mail: c.w.moran@umail.ucc.ie

have been shown to produce 'profound' transformation in practices, meanings, and discourses concerning food.

In this chapter, we aim to advance the discussion on AFNs by focusing on how digitalisation reconfigures consumers' learning and engagement with food. While the digitalisation of AFNs has started to attract increasing scholarly attention (Cui, 2014; Bos & Owen, 2016), empirical research in this area has focused mainly on the potential of digitalisation to improve the organisational efficiency of alternative provisioning (De Bernardi et al., 2019). On the producers' side, scholars have highlighted how digital platforms can facilitate direct connections between small farmers and their customers, and among producers themselves, thus potentially enabling the construction of new organisations in the form of alternative digital food networks (in efforts to 'scale-up' or 'out', i.e. replicate) and grassroots local food initiatives (Bos & Owen, 2016; Schneider et al., 2019; Lewis, 2020). On the consumers' side, the digitalisation of AFNs has been seen as a way to overcome obstacles relating to time, effort, and lack of physical space, making engagement and access to 'good food' more practicable (Dal Gobbo et al., 2022; Fuentes & Fuentes, 2022).

By departing from the assumption that digitalisation represents a specific form of intermediation and by drawing on in-depth interviews with consumers involved in different types of physical and digital AFNs in Ireland, Germany, and Italy, this chapter seeks to throw light on how the organisational and operational structures of offline and online alternative provisioning services (their design and ways of operating) affect consumers' skills and knowledge about food. In other words, our main goal is to better understand how digitalisation affects the 'learning by doing' which has often been seen as an important feature of alternative food procurement networks.

We begin by addressing the issue of deskilling in food consumption and production and highlighting how enabling and promoting peer-to-peer exchange through alternative provisioning platforms can represent important opportunities for consumers to regain food knowledge and competencies (Savarese et al., 2020). The chapter then presents the characteristics of the AFNs studied within the PLATEFORM project in Ireland, Germany, and Italy. The analysis of 83 in-depth interviews conducted with consumers involved in both offline and online schemes will then help to understand whether and to what extent physical and digital alternative provisioning services are shaped to accommodate open and ongoing learning processes among consumers. In particular, by investigating the daily practices of people involved in these networks, our analysis will help to grasp how their different organisational and operational structures encourage processes of learning and reskilling. While wrapping up the main findings, the concluding section will discuss the potential of new digital platforms to extend the reach of physical AFNs by providing new learning spaces.

## 2 Consumer Deskilling and Reskilling in the Age of an Increasingly Industrialised and Corporate Food System

Food has increasingly become a manufactured product. Thus, preparing a family meal with raw and seasonally appropriate ingredients has become increasingly less common (Warde et al., 2007; Plessz & Étilé, 2019). As Jaffe and Gertler (2006) argue, due to a constantly changing, highly technological, and geographically distant food system, consumers are becoming less skilled. This 'deskilling' has several consequences. Firstly, outsourcing the laborious and skilful task of food preparation from the household leads to a lack of opportunities for learning:

> *Food production has traditionally been learned through apprenticeship, with children learning first-hand while their mothers cook. These skills are sentient, practical, and in some senses non-discursive forms of consciousness, with the learner acquiring a knack, or a feel, that comes with the continual engagement with the physical and sensual qualities of food. (Ibid., p. 147)*

Furthermore, the growing distance between production and consumption leads to a mutual lack of knowledge between producers and consumers (Jaffe & Gertler, 2006). Consequently, many consumers have not learned to recognise high-quality and high-value produce. Producers, on the other hand, learn about consumers' needs only through the processors of food and food manufacturers and retailers. As a result, there is only fragmentary knowledge on both sides about either cultivation or working conditions or respective needs and requirements.

As observed, deskilling has happened gradually, with each generation of food provisioners having fewer skills than the one before (Jaffe, 2015). As manufactured food was introduced, consumers' cooking skills started to fade giving rise to major shifts in food practices (Colatruglio & Slater, 2014). Although it is commonly believed that convenience and processed foods were developed in response to social changes, research has shown that, in fact, the growth of the food industry drove the changes rather than their being a response to consumer demand (Jaffe, 2015). Looking at the situation in the United States since the beginning of the twentieth century, for example, Shapiro (1986, pp. 79-80) highlights how 'decade by decade, as proponents of packaged-food cuisine worked tirelessly to make it the centre of American cookery, they met less resistance from women who lacked the taste memory and the skills their mothers or grandmothers had brought to the kitchen'. As such, the decrease in the knowledge and expertise of traditional cooking skills has resulted in a greater dependence on pre-packaged and commercially produced food, ultimately leading to the expansion of large-scale agriculture and the integration of market economies into areas that were previously outside of the commercial system.

As Jaffe and Gertler (2006) contend, developments in the food system must be seen as the result of an evolving two-way relationship between new technologies and/or food products and consumer skills. As these authors put it, the more technology evolves, the more consumers become deskilled, and, in turn, the more they need new technology. In other words, when knowledge or skill is lost, it is replaced by a corresponding technical innovation; a process that, in their view, has had two important effects on the development of the food system: (1) it has produced new consumers who devote a minimum of time and effort to food, in turn generating new consumption patterns and new markets for food products and technologies, and (2) it has established food as an arena of fad, fashion, and obsolescence (Jaffe, 2015, p. 352). To illustrate the complexity of knowledge and skills involved in the food supply, Jaffe (2015) identifies ten types of deskilling ranging from loss of the ability to use human senses and competences in separating unwanted from nutritious foods, to lack of knowledge and control over food choices and how the food system affects the rest of society (Box 1).

Issues of food literacy and reskilling today have become of paramount importance, as 'the relationship people have with food is becoming even more disordered as obesity and diet-related chronic disease rates soar. Messages emphasising diet and nutrition are numerous and conflicting from public health authorities proposing healthy eating and for food companies marketing their vast and diverse product for profit' (Colatruglio & Slater, 2014, p. 37; see also Dal Gobbo in this volume). Still, the McDonaldisation literature (e.g. see Ritzer, 2000) stresses that people will

**Box 1 Jaffe's Ten Types of Deskilling (Jaffe, 2015, pp. 353–354)**

1. Sense–perceptual: Loss of knowledge in engaging the senses. E.g. Touching or tasting a product to see if it has expired.
2. Motor–manual: Loss of skills that use the body. For example, cutting, whipping and mixing.
3. Practical–material: Loss of familiarity with ingredients and materials used in food procurement.
4. Processual-systems: The idea of human beings becoming "machine minders" who know "only one's own part of the division of labour".
5. McDonaldization: Standardisation and efficiency become the dominant criteria in process and consumption more so than factors based on quality.
6. Intellectual: Includes loss of knowledge about ingredients, dishes, nutrition, label reading and food safety, which is related to professionalisation in the food system whereby "proper knowledge is that which is held and known by proper knowers, such as nutritionists or dieticians".
7. Conceptual: Loss of creativity in food provisioning and preparation. E.g. Having difficulty in creating a meal from existing ingredients.
8. Organisational: Deficit of knowledge required to plan meals in the short and long term and to shop in an appropriate and economical manner.
9. Social: Loss of knowledge in food-related practices that maintain and protect sociality and culture. Such skills "link food provisioning to the broader sphere of social life and are interactive and involve knowing what is appropriate in what context, when, and with whom".
10. Civic: Loss of knowledge about "how food choices and the food system affect the rest of society".

increasingly want easier, faster, and cheaper access to everything, therefore continuing the progressive deskilling that will make it even more difficult for most consumers to recognise and respond to challenges presented by an increasingly industrialised and corporate food system (Schlosser, 2012). In this context, the spread of AFNs may provide an appropriate and timely opportunity for effective resistance around reskilling and food sovereignty.

By shortening food supply chains, AFNs have made local food more accessible whilst (re)building meaningful relationships of mutual support and trust between producers and consumers (Grasseni, 2013). A growing number of empirical studies have underlined how these initiatives act as learning catalysts amongst their members subsequently leading to improved sustainability practices in the food system. In-person interaction at farmers' markets (Milestad et al., 2010) and direct involvement in CSAs (Cox et al., 2008; Feagan & Henderson, 2009) or Solidarity Purchase Groups (Forno et al., 2015; Signori & Forno, 2019) can facilitate two-way communication between consumers and food producers and have significant repercussions for the food system as it increases participants' awareness of broader sustainability issues (Forssell & Lankoski, 2015). Participating in AFNs can also be an entry point to learning useful cooking and preservation practices. As observed by Brunori et al. (2012), through both face-to-face and online (mainly e-mail) communication exchanges, members of Solidarity Purchase Groups can understand the link between food consumption and broader sustainability issues. Their horizontal organisational structure also encourages learning amongst peers who have the opportunity to (re)acquire skills and knowledge about how to cook raw foods or 'hard' vegetables, therefore to (re)acquire the necessary skills which help to overcome the so-called attitude-behaviour gap (Carrigan & Attalla, 2001).

Along these lines, a recent study in New Zealand found that being a CSA member can be a significant factor in promoting consumer education and behaviour change as 'deep involvement allows consumers to become ambassadors for processes of change in the regions where they live, and to share their experiences in a continuous "learning by doing" process' (Savarese et al., 2020, p. 11). Further research has also suggested that farmers/producers can play an important role in facilitating learning amongst AFN members via informal and incidental knowledge sharing (Etmanski & Kajzer Mitchell, 2017). Put differently, participating in AFNs—while favouring direct encounters between consumers and producers—appears to facilitate learning on both sides of the food chain and at different levels. This occurs both on a concrete, practical, and everyday level in relation to household food practices and also on a higher, broader level in relation to awareness and concern for sustainability-related issues in local and global food systems.

These findings are of utmost importance as learning and awareness regarding sustainability issues has been identified as a significant antecedent of an increase in sustainable food practices (Kloppenburg et al., 1996; Sundkvist et al., 2005). By linking different actors with different forms of knowledge and expertise in situated actions, AFNs—as a specific type of communities of practice (Wenger, 1998)—can in fact play an important role in fostering social learning, that is, a type of learning generated through observing the behaviour of others which has been identified as an important tool for 'responding to sustainability challenges and conceiving of new pathways for transition' (Slater & Robinson, 2020, p. 13).

Although research on online provisioning food platforms is still scarce, recent studies on digitally supported AFNs have highlighted not only their potential to make the shopping procedure easier and quicker, simplify logistics and the organisation of payments, etc. but also their role in facilitating and enabling learning and engagement among members/customers in the context of sustainable food consumption practices (De Bernardi et al., 2019; Savarese et al., 2020; Veen & DAmico, 2019). In a similar vein, a study conducted in Italy suggests that digital and physical (on-site) knowledge sharing amongst customers of a Food Assembly platform can promote positive change towards more sustainable purchasing behaviours (De Bernardi et al., 2019).

In the following sections, we will examine the different ways in which learning occurs within both offline and online alternative food networks (AFNs) while considering the ten types of deskilling identified by Jaffe (2015). We will draw on data collected during the PLATEFORMS project to investigate the extent to which these new supply networks enable individuals to regain lost knowledge and skills related to food. Our primary focus will be on how digitisation transforms the traditional 'learning by doing' process and how new supply networks facilitate reskilling among members/consumers. By investigating the characteristics of alternative supply networks, we hope to identify innovative mechanisms that can be transferred to other food supply channels—such as school and public canteen supplier networks—to promote more widespread sustainable consumption practices.

## 3 Cases Description and Empirical Material

Our analysis builds on 83 in-depth interviews conducted in Ireland (20 interviews), Germany (30), and Italy (33) with people purchasing their food through 10 different physical and digital AFNs. Interviewees displayed a fair degree of variation in terms of age, education, and socio-economic status. The ten selected platforms are characterised by different organisational and operational structures, which we divided into four models by considering their ownership, level of consumer involvement, participation, and interaction among consumers and between consumers and farmers. We have labelled these four models: (1) grassroots physical food platforms; (2) grassroots food digital platforms; (3) food assembly platforms; (4) digital box schemes.

### *Grassroots Physical Food Platforms (CSA, GAS)*

Grassroots physical food platforms can be considered as one organisational form within the contemporary civic agriculture movement, and they work by connecting consumers and producers in a physical space. In the three countries considered, 18 interviews were conducted with members of CSAs—ten in Ireland and eight in Germany. For the Italian case, 11 interviews were conducted with members of Solidarity Purchase Groups (Gruppi di Acquisto Solidale—GAS). Although they have some organisational differences, CSAs and GASs share several common characteristics and they both aim to create direct connections between producers and consumers.[1] The general goal of such networks is to have producers and consumers enter into market exchanges as equal parties by agreeing on fair prices and fair wages. Members of such organisations are often active citizens engaged in several activities, such as self-organised food collectives, dumpster diving, or local food councils aiming to shape local food politics.

[1] A CSA, or Community Supported Agriculture, is a model where consumers become members of a farm, and in exchange for a fee or subscription, receive regular deliveries or shares of the farm's produce. This model is often seen as a way to support local agriculture, promote sustainable farming practices, and build community connections. On the other hand, Solidarity Purchasing Groups (GASs) are a type of consumer collective that emerged in Italy in the 1990s. By pooling their resources and buying power, these groups aim to create a more equitable and sustainable food system. While CSAs and GASs share a similar goal, they differ in their focus and structure: CSAs support a specific farm and its operation, while GASs focus more on collective purchasing power and building a network of sustainable suppliers.

**Fig. 1** Grassroots food platforms

Exchanges in CSAs and GASs are based on values of solidarity, transparency, environmental stewardship, and producer-consumer relationships. Being part of a CSA or a GAS involves attending regular meetings of members to discuss the products and politics of the group. The participating households pick up their weekly harvest at multiple distribution points, which can be a farm or any space within a private or public building. Both CSAs and GASs usually make very little use of digital tools. At most they offer a website with an overview which is not interactive or a simple chat group to coordinate the pick-ups (Fig. 1).

### *Grassroots Digital Food Platforms*

The label 'grassroots online food platform' has been applied to digital non-business infrastructures that facilitate networking among consumers and between consumer groups and producers. These digital platforms provide a centralised IT service and thus a shared virtual space for members. Two types of organisations are considered in Germany and Italy with a total of 16 interviews. The German organisation 'Foodsharing' (eight interviews) is a volunteer-run, self-managed organisation founded in 2012 that rescues surplus food from different companies, for example, bakeries, supermarkets, hotels. The members (food savers) pick up the products that would otherwise be thrown away and are then responsible for the (non-commercial) distribution of the food. Food savers use the online platform to coordinate the food rescues and food-sharing activities and also to communicate with each other, for example, about recipes but also about general politics of the network. Furthermore, there are regular (online) meetings in each neighbourhood where food savers meet to discuss different aspects, for example, new food shops where surplus can be saved. ‘Buonmercato’ (eight interviews) is an Italian association based in Corsico (Milan periphery) founded in 2010 directly out of the GAS experience. As with Foodsharing, Buonmercato is a volunteer-run, self-managed organisation. The centralised IT service allows members to place their orders online. Once orders are made, the volunteers aggregate them and send them to the producers. Producers directly bring or send the products to Buonmercato’s headquarters, and members pick up their orders at the association’s offices on a weekly basis. Although the way Buonmercato operates is reminiscent of food assembly platforms, it is the members themselves who choose the producers, establishing a direct relationship with them (Fig. 2).

### *Food Assembly Platforms*

Food assembly platforms (FA) are a centralised IT service which can be described as a ‘digital farmer’s markets’ where local producers/suppliers and customers engage in transactions digitally. It therefore offers an innovative type of direct selling mechanism between local producers/sellers and consumers in the local community. Across the three countries considered, 29 interviews were conducted with FA customers: 10 in Ireland, 7 in Germany, and 12 in Italy. Within these digital platforms, producers present themselves (sometimes with family, stock, and employees) and their products. Every week the consumers choose and pay for their products online according to the (varying) offer of the FA. Each FA has distinct times when the products can be picked by. Interestingly, the pickup points are at diverse locations like organic bakeries or tea shops. Hence, the digital platform is embedded in an already existing analogue local and community (food) infrastructure. Platform suppliers vary from direct sale farms selling fruit, vegetables, meat, and honey to businesses operating

**Fig. 2** Grassroots digital food platforms

their own physical and/or online shops while also offering their products through the platform (e.g. butchers, coffee roasters, and cheesemongers). The Irish FA[2] was founded in 2018, while German and Italian food assemblies considered in this study (Marktschwärmer and Alveare che dice sì) were both launched in 2014 and operate on the same IT service developed by a French trading company (Ruche and Equanum SAS) that is inspired by the principles of the collaborative economy. This means that the service aims to facilitate more direct and transparent connections between producers, processors, and consumers, with the aim of redistributing the transactions between these groups more equitably. These IT services are thus intended for the marketing of agribusiness production and agribusiness processing, with managers being the interested intermediaries (Fig. 3).

[2] The real name of the food assembly in the Irish case has been anonymised in order to adhere to the requirements set out in the ethical approval obtained through the university.

**Fig. 3** Food assembly platforms

### *Digital Box Schemes*

With the label 'digital box schemes' we have aggregated online food provisioning schemes offering fresh food from local producers to be delivered directly to the homes of the consumers. In these cases, the company owning the digital platform manages all the logistical aspects, from the technical infrastructure to the collection, packaging, and delivery. Within the sample of this study there are two digital box schemes making a total of nine interviews. The regional boxes in Germany (seven interviews) are a concept offered by more than 40 agricultural enterprises all over Germany. There are currently six different enterprises in and around Berlin, among others the Apfeltraumkiste, the Märkische Kiste, and the Ökodorf Brodowin. Most of them have a longer history of selling and distributing their produce directly to consumers, one of the largest platforms (Ökodorf Brodowin) consists of various farmers who have retailed their produce directly since 1995, for example. The boxes are delivered directly to homes and the content is usually unknown to the consumer beforehand. Depending on the enterprise and what it offers, consumers may be able to change some aspects of their box online, and sometimes even additionally order from a full range of organic foods beyond what is offered regionally. Thus, even though the boxes foster a direct producer-consumer relationship, communication is very limited and takes place mainly digitally. The Italian case (two interviews), Cortilia S.p.A., is an e-commerce business delivering fresh food products to homes across the North of Italy. It was founded in 2011 as an online fruit and vegetable market. The company chooses artisanal, small-to-medium, producers. Prospective consumers find products online and the company delivers them directly to their homes in refrigerated vans. Products can be ordered *à la carte* or customers can subscribe to a specific box scheme (Fig. 4).

While analysing our data, particular attention was paid to how the organisational and operational structures of offline and online alternative provisioning services (their design and ways of operating) work to accommodate open and ongoing learning processes and to how digitalisation reconfigures learning and engagement with food. Interviews were first transcribed and translated to English for a uniform analysis, then the

**Fig. 4** Digital box schemes

analysis proceeded through collaborative thematic, and a mix of inductive and deductive, coding. Data and researcher triangulation (Patton, 2002) allowed us to distinguish four main areas of learning through which we have structured the analysis that follows.

## 4 (Re)acquiring Knowledge and Skills Through Learning by (Alternative) Provisioning

As discussed in the previous pages, new supply networks allow consumers and producers to be reconnected. By re-arranging the materiality of food through direct exchange, these new supply channels have been recognised as playing an important role in the revitalisation of local economies. In this section, we aim to explore whether and how belonging to these networks increases the ability of individuals to understand the intricate interdependence between environmental, social, and economic issues underlying the contemporary food system. We are interested in exploring whether the digitisation of alternative food networks (AFNs) reconfigures consumers' learning and engagement with food and whether and how this occurs differently to grassroots physical food platforms.

The collected interviews contain numerous references to how purchasing through new supply networks is perceived by the interviewees as connected to a higher level of knowledge and comprehension regarding how the food system works.

Overall, the data indicate that respondents associate participation in these food platforms with increased knowledge and skills in four different learning areas:

(a) About farms and fields: learning about cultivation and production
(a) About apples and kale: learning about products (regional/seasonal/new products)
(b) About pickling and jam: learning about preparation and storage
(c) About politics and products: learning about how food systems work (food wastage, farmers' exploitation, soil degradation, $CO_2$ emissions, etc.)

In the following sections, we will examine whether, how, and to what extent the various food provisioning platforms described above facilitate the acquisition of knowledge and skills related to these different areas of food.

### *About Farms and Fields: Learning About Cultivation and Production*

The pursuit of food that is both delicious and beneficial for personal health and the health of the planet is frequently the primary motivation for exploring alternative supply channels, as demonstrated in other chapters of this anthology (see Dal Gobbo as well as Brettin et al. in this volume). At the same time, it could also be the starting point for the acquisition of new knowledge and skills that question standardisation and efficiency as the dominant criteria in processing and consumption ('McDonaldization' in Jaffe's terminology) and to reacquire expertise on the multiple dimensions of food quality. CSAs, Solidarity Purchasing Groups, and grassroots digital food platforms usually require their members to share some work with farmers such as harvesting and delivering. All these activities are judged by informants involved in these food networks to be integral to directly experiencing where food comes from and how it can be cultivated more ecologically.

According to VP, a member of a German CSA, participating in the harvesting organised by her group was a very positive experience. As she said, the CSA she is affiliated with requires members to help a few times a year with different tasks. One of the benefits of taking part in these activities that VP mentions is that they show where food actually comes from. In particular, she pointed out how learning by doing gives participants a better appreciation for the work that goes into producing the food they eat, thus facilitating the acquisition of knowledge regarding the other part [the producer's] of the division of labour (reskilling in processual systems, as defined by Jaffe):

> *We participated in the harvest of the CSA. There is a bit of a requirement for our CSA that you help out three times a year or so somewhere with the harvesting or delivering or something. And we went to Potsdam and harvested carrots and potatoes last autumn. I mean, for one thing, it was a really nice day, and it was a lot of fun. But on the other hand, it was also really nice to see where our food actually comes from.* (VP, female, CSA, Germany)

Despite the time commitment, peer learning in common activities is also generally considered important for obtaining first-hand information, which respondents consider crucial for increasing their autonomy in making purchasing choices. During the interview, ET, a member of an Italian GAS, reports that people often ask her why she attends so many meetings to decide what to buy. For her, the opportunity to discuss food issues with her group members is an important way to make more informed food choices:

> *People often ask me why I'm attending all these meetings, and I am using all this time … however, I think it all comes down to the fact that I get some interesting information that helps me do some thinking about the food which then helps me to devote more attention and to decide not to accept everything.* (ET, female, GAS, Italy)

Collaboration and cooperation appear therefore to provide a meaningful opportunity for consumers to develop a deeper understanding of their purchases. Especially when individuals interact face to face with producers and collaborate directly with them, they can expand their knowledge and engagement beyond just food safety. This type of collaboration is particularly prevalent in grassroots alternative food networks.

However, if digital food platforms offer limited opportunities for face-to-face interaction, our data reveal that the exchange between producers and consumers facilitated by these platforms can still foster a greater understanding of farms and fields than that typically offered by traditional supermarkets. Consumers who purchase food from these platforms appear, in fact, to develop a better understanding of the origins of their food and the processes involved in its production. Such was the case with RM, a member of a German FA, who highlights in his interview that purchasing through food assembly platforms increases the ability to discern when a product is fresh—a distinction that he maintains is often difficult to make in supermarkets. He suggests that this knowledge enables consumers to make more informed consumption choices, leading to a more conscious approach to food purchasing:

> *[When buying in an FA] you know when it's fresh, you also know why things don't work out or what problems there are. These are all things that you don't even notice in the supermarket because it is somehow buffered by three different middlemen or a wholesaler and another dealer and the supermarket itself. And you also learn more about agriculture or the production process and perhaps eat more consciously as a result.* (RM, male, Food Assembly, Germany)

It is noteworthy that when comparing the narratives of those who participate in the four different types of food provisioning platforms examined, it becomes apparent that purchasing from food assembly platforms or through food box schemes generally only tends to increase consumers' knowledge about product characteristics. These platforms rarely provide consumers with an opportunity to acquire more information about the functioning of the entire food chain, which is more frequently the case in grassroots alternative food networks. Despite this limitation, it appears that purchasing food from online food platforms can enhance consumers' understanding of the sensory aspects of the food they consume, such as taste, texture, smell, and appearance, referred to by Jaffe as 'sense perceptual'. Respondents who purchase through these networks describe a different shopping experience than that provided by traditional procurement channels.

Nonetheless, it is important to note that not all digital food procurement platforms work in the same way. For instance, in the German case of regional fruit and vegetable boxes, consumers are often invited to get in touch directly or to visit the farm. However, the interviewees themselves testify that this offer is seldom taken up unless there are specific problems, such as when the crates contain different quantities of produce than expected, perhaps because an insufficient harvest means all consumer demands cannot be met. This is exemplified in the case of AB:

> *Then I called them and asked: 'How did that come about?' And then they said that when they were packing, they realised they didn't have enough peppers. Because that can happen when packing boxes, someone hasn't harvested enough.* (SG, female, digital food box, Germany)

### *About Apples and Kale: Learning About Products (Regional/Seasonal/New Products)*

As discussed in the introduction of this chapter, the trend towards an industrialised and corporate food system has led to a decline in consumer knowledge about the geography, seasonality, and availability of food. This phenomenon has been described as 'intellectual de-skilling' by Jaffe (2015). In the collected interviews, there are many references to the fact that through alternative supply channels, consumers can recover these skills.

When asked about which foods she has started using more of since joining the CSA, IC, a member of an Irish CSA, named kale: *'Obviously kale, I wouldn't have eaten kale before'*. IC went on to explain how she has also learned to make a kale pesto in order to use up the vegetable which was abundant at that time of year:

> *We've so much kale, but it's lovely just to cook a pot of it, and when it's done you take off your water and save that for tomorrow's soup, and then make a really nice dressing over the kale. I got a recipe from a neighbour, it's lemon juice, olive oil, garlic and salt, and you can blend it or else just give it a good mix and just drizzle that over your kale and it's absolutely delicious.* (IC, female, CSA, Ireland)

The discovery of new products through participation in AFNs was commonly reported by interviewees. Some of them even started gathering ingredients from nature, as reported by AL, a member of an Irish CSA, who started foraging to supplement the food available through her own food network:

> *Not everybody but a lot of people living here are also in that—going around and foraging and so there's always lots of tips and advice and people to share information with.* (AL, female, CSA, Ireland)

Participation in these food networks also appears to not only expand knowledge of new local foods but also encourage the consumption of seasonal and less commonly consumed fruit and vegetables.

In the case of LV, a member of an Italian GAS, learning about the seasonality of products significantly influences her diet, as gaining new knowledge motivated her to make more conscious choices about what to eat. Her example demonstrates how alternative food networks can empower individuals to make more heedful decisions about their food consumption and, in so doing, contribute to a more sustainable and healthier food system:

> *I learned the cycle of the seasons, I used to eat zucchini even in December, I found out that zucchini is not a winter vegetable but is definitely more of a late spring-summer product. So having this kind of information, I changed my diet.* (LV, female, GAS, Italy)

Food geography and seasonality are particularly recurring themes in the interviews conducted among Italian GAS members, who, compared to the Irish and German interviewees, generally demonstrate greater knowledge of ingredients, food varieties, dishes, and nutrition. However, some interviewees point out that consuming only seasonal foods can be a struggle, especially when there are children in the family.

Interestingly, our interviews showed that people sourcing through FAs and GASs often have a recipe in mind and then purchase the necessary ingredients through their alternative food network, while CSA and box scheme members seem to work more with available products:

> *Sometimes things come, so as I said, you don't choose every food there. And you take it a bit as it comes. And I think that's good because, as I said, these are things that are growing right now and that are somehow on offer. And I always think a little bit: 'That's what the soil is offering now, and that's what I'll eat.'* (SG, female, digital food box, Germany)

This finding suggests that FA and GAS customers may have a higher degree of food autonomy and a stronger focus on personal preference in their food choices, while CSA and box scheme members are more willing to adapt their meals to what is seasonally and locally available. This could be due to the fact that in CSA and box schemes, consumers receive a predetermined selection of produce, whereas in FAs and GASs, they have more control over what they purchase.

After analysing interviews collected from members of new physical and digital provisioning channels, it is evident that there is a significant difference in how knowledge is exchanged within these networks. In CSA and GAS groups, there is a lot of sharing of products and recipes, both in person and through mailing lists and social networks such as WhatsApp or Telegram. By contrast, customers of FAs tend to learn more independently through using searching on Google or YouTube:

> *I googled turnips actually, because I'd gotten small turnips in the veggie box.* (JE, female, Food Assembly, Ireland)

> *I was making something and I wanted to put pickled onions in it so I just googled how to make pickled onions.* (LF, female, Food Assembly, Ireland)

> *There are a couple of dishes that I've gotten from cooking shows. I like cooking shows and I like watching things on YouTube and following their recipes.* (ST, female, Food Assembly, Ireland)

While individuals who utilise grassroots physical platforms often learn from each other, the convenience of accessing information online seems to complement the lack of face-to-face interaction in more digitally mediated food platforms.

### *About Pickling and Jam: Learning About Preparation and Storage*

As our interviews revealed, both physical and digital food provisioning networks offer valuable learning and apprenticeship opportunities also in areas such as food preparation and preservation (motor-manual, practical-material, conceptual, and organisational reskilling according to Jaffe's categorisation). In particular, many of our informants reported how being a member of a grassroots physical food platform, such as a CSA or a GAS, has helped them learn ways to preserve food such as canning, pickling, fermenting, and freezing.

For example, in her interview, JS, a member of a German CSA, shares how being a part of a CSA group allows the exchange of knowledge and tips among members. In this context, she learnt how to ferment from another member, an experience she valued very much as it allowed her to acquire useful skills that she might not have otherwise gained:

> *In the group, you also have a nice exchange and give each other tips and, I don't know, there are people who can ferment and then you do it together. So, these are useful skills that I would have had to go to a lot of trouble to acquire or something that I would never have done.* (JS, female, CSA, Germany)

During the interviews, respondents often complained that they lacked adequate culinary skills, attributing this to a loss of familiarity with ingredients and the types of technologies and materials used for food procurement. Although many admit their limited skills in the kitchen, there seems to be a high degree of recognition among respondents of the benefits they derive from sharing knowledge and recipes with each other, as described by EA and ND, both members of the Irish CSA:

> *Our neighbour does incredible preservation for the winter ... She's always sending out recipes with the harvest report—that's another great thing.* (EA, female, CSA, Ireland)

> *I wouldn't be great at cooking. I can do various basic things, but I am learning. When you are in the coach house there might be somebody in there, and you say: 'What do you do with that?' and they give you a recipe. They tell you about the likes of this. I got that off somebody else, just talking to someone: 'Try it this way and try it that way'.* (ND, male, CSA, Ireland)

Confirming the findings of previous research (Savarese et al., 2020; Etmanski & Kajzer Mitchell, 2017; De Bernardi et al., 2019), AFNs appear crucial in promoting food knowledge, as they are spaces where people can share what they know and help each other regain the skills needed to make more sustainable purchases. Participation in AFNs allows individuals to regain and learn the skills needed for proper food management.

In order to address the issue of deskilling in food provisioning and preparation, many groups digitally share new provisioning channels that include recipe suggestions and cooking tips along with their products. As SG, who buys her fruit and vegetable through one digital food box in Germany, said in her interview:

> *Now with a lot of things you don't know right away instinctively what to do with them. But everything is described. And then you also learn that a little bit. That's quite cool. Because of course, as soon as the things are harvested, they lose their nutrients and so on.* (SG, female, digital food box, Germany)

The online resources provided by these services—in the form of printed instructions or posted on their websites—are considered particularly useful by respondents, especially by those who are not familiar with certain vegetables or their preparation:

> *I liked to experiment with the app designed with a kit that allows you to create dishes by giving you also an idea from a nutritional point of view.* (SP, female, digital food box, Italy)

It seems that by receiving a variety of fresh, seasonal produce each week, users are encouraged to cook with ingredients they might not otherwise have tried and they find creative ways to use up everything in the box. This leads to a sense of excitement and adventure in the kitchen. As vegetable box customers in Germany and Italy report, digital food box services are appreciated not only for their convenience but also because they expand users' culinary knowledge:

> *With the box, I just get it presented in front of me, and then I'm just happy that I can/ Then I first look: What am I going to do with it now? I don't know what I can cook with it now, what I can do with it at all. And then it's just the other way around. Otherwise, you have this recipe in your head and you buy some-*

*thing for this, and this way you are presented with things and then you look for a recipe. And that's actually always quite nice.* (GK, female, digital food box, Germany)

*With Cortilia, there's this fancy thing that they have a lot of weird vegetables, some slightly bizarre variations. Last time there was this catalogna, a miniature catalogna. It ended up being fun. In these months we do the savoury pie once a week, well every week we change we take something with what's offered. Let's say I'm someone who for years ate what the supermarket offered, now I have a different kind of nutrition, something else is shown and so we try.* (GA, male, digital food box, Italy)

All in all, our interviews suggest that being part of alternative provisioning, both physical and digital, greatly stimulates creativity in food preparation, helping people to regain competence in how to put together meals with existing ingredients or adapt recipes to those available.

### *About Politics and Products: Learning About How Food Systems Work*

As previous research has shown (Savarese et al., 2020; Etmanski & Kajzer Mitchell, 2017), AFNs are valuable contexts for individuals to develop a better understanding of how food choices and the food system impact society, thereby enhancing their civic skills. Many interviewees also emphasised the importance of this capacity. For example, EA, a member of an Irish community-supported agriculture group, noted that participating in a CSA has increased her awareness of sustainable and healthy food production practices, which are crucial for both human and environmental health. In her interview, she points out how conventional food production and pesticide use on the land can lead to the poisoning of our food. Being part of a community-supported agriculture group has had the effect of making her more careful about what she and her family eat, encouraging her to consider the impact of her daily choices on the environment:

*Now I'm a member of the community farm, I can see how well it can be done and how important it is, what goes onto the land. I just feel like general agribusiness type production, or general standard production, where everything gets sprayed and waxed and basically the ground gets poisoned to keep weeds at bay, I don't want that going into my system and my child's system, I really don't. I'm much more aware of that now than I used to be.* (EA, female, CSA, Ireland)

During her interview, EA was also keen to emphasise how her experience in the Irish CSA not only helped her become more aware of the importance of sustainable and healthy food production but also made her realise the importance of educating others about it. In this regard, she recognises that AFNs can play a crucial role in promoting food security and that it is essential to increase people's awareness of the challenges facing the global food system. As she stated in the interview:

> *I think one of our big things with the community farm is just to educate other people on how important this is and how fundamental it is to future food security and [that] to be able to access really healthy food is extremely important for everyone.* (EA, female, CSA, Ireland)

The possibility of exchanging knowledge and ideas and thus empowering and encouraging individuals to make more informed choices about their food consumption was clearly more evident among participants in grassroots food platforms than in digital ones. Indeed, in face-to-face situations, it is common for members to remind each other of the 'alternativeness' of alternative food networks. During discussions, there may always be someone who reminds everyone why it is important to support local producers, reduce food waste, and promote sustainable agricultural practices, thus helping to link the food supply to the broader sphere of social life, indicating what is appropriate in what context, when and with whom (social reskilling).

Once involved in these initiatives, consumers may also be introduced not only to new produce but also to other projects. This is what happened to LV, a member of a GAS, who recounts in her interview how—through buying fruit and vegetables for her family—she became aware of, and subsequently started to support with her purchases, a social project aimed at offering prisoners the opportunity to acquire new skills and work experience:

> *[…] then during Christmas there are the products of the prison economy, all produced in the prisons of Italy, sweets mostly […] this is a very nice project carried out in women's and men's prisons.* (LV, female, GAS, Italy)

Learning about the problems associated with the industrial food production system is particularly evident among the participants of grassroots physical food platforms. This is not to say, however, that certain knowledge and sensitivities are not present among consumers who buy food through food assemblies or box schemes, as was clear in several interviews, such as in the cases of ML and SR, members of an Italian and a German food assembly, respectively. They were keen to emphasise how the choice to buy through an alternative supply channel is aimed at giving less money to large food chains and at supporting small producers instead:

> *Yes, to me anyway, the idea of giving less money to the big chains is stuff that I like, in the sense that I like the idea of giving a helping hand to small producers. To me this thing that they are disappearing or that life is difficult for them bothers me.* (ML, female, Food Assembly, Italy)

> *But I know how much work is involved and, phew! what quantities of cream are needed. And I have also discussed this with the people there, with [name], and she says: 'Actually, we make the butter at cost price.' They don't even earn anything from it. So that shows once again what the industrial production of food actually does, also in terms of prices.* (SR, female, Food Assembly, Germany)

Ultimately, the collected interviews suggest that the absence of physical encounters in online alternative procurement platforms diminishes the possibility of accidental knowledge sharing (Etmanski & Kajzer Mitchell, 2017), ultimately affecting the broad educational capacity to consider all the multiple dimensions of food quality. Only in the case of digital platforms that maintain a dual physical and digital dimension does the self-educational capacity of grassroots AFNs seem to combine with the greater convenience of online shopping. This seems to be so, for example, in the case of the grassroots digital platform, Buonmercato. In this regard, the experience of VS is interesting. In her interviews, she recounts how the time she spent with Buonmercato members contributed to increasing her awareness and motivation to change her eating habits, making her consider reducing her meat consumption, both for the sake of less pollution and in order not to indirectly contribute to the exploitation of animals.

> *Spending time with the people of Buonmercato has contributed to our increased awareness and motivation to change our eating habits. Through our conversations with them, we have become more informed about food and sustainability topics.* (VS, female, Buonmercato, Italy)

## 5 Discussion and Conclusion

This chapter aimed to contribute to the ongoing debate on the potential role of AFNs, both physical and digital, in facilitating consumer learning and reskilling. Indeed, as manufactured food was introduced, consumers' skills and competence in provisioning started to disappear, giving rise to major shifts in food practices. As observed above, the multiple dimensions of quality related to food include the contributions that a well-chosen diet can make to human health, planetary sustainability, and community economic development. As argued by Jaffe and Gertler (2006), consumers who are unaware of, or unconcerned about, such dimensions of quality are less likely to demand better food or foodstuffs. Moreover, the consequences of consumer deskilling are equally serious for farmers and farming. The growing distance between producer and consumer means that farmers/producers receive information on 'what the consumer demands, only via food processors, manufacturers, and retailers. This filter biases information flowing in both directions. Those who must shop to eat end up with a very cloudy or partial picture of growing conditions and production practices as well as farm economics and farm structure' (Jaffe & Gertler, 2006, p. 146). Thus, if we want to move towards a more sustainable food system, finding ways to 'reskill' consumers appears to be of paramount importance.

In this regard, our findings suggest that new supply networks can promote more informed choices and purchasing behaviour (De Bernardi et al., 2019) since they help people regain a certain level of autonomy and control over their consumption choices and actions. Consumer learning and reskilling across various platforms encompasses the complete range of abilities identified by Jaffe (2015). While it is evident that certain food platforms, such as grassroots physical platforms, yield stronger and more comprehensive learning outcomes, the analysis highlights that shopping through new digital supply channels offers advantages that extend beyond convenience in accessing food.

Through virtually mediated exchange between producers and consumers, individuals are able to regain a deeper understanding of the sensory aspects of the food they consume, including taste, texture, smell, and appearance (sense perceptual reskilling). Additionally, these channels induce consumers to approach meal planning and cooking in a more innovative and creative manner (organisational reskilling). As it turns out, interestingly, this is especially true for people who buy food through digital food box programmes. Similarly to what happens in grassroots physical food platforms (especially CSAs), it was possible to observe how receiving a variety of fresh, seasonal produce each week encourages these people to cook with ingredients they might not otherwise have tried and to find creative ways to use up everything in the box (practical-material, motor-manual, and conceptual reskilling).

As such, along with providing consumers with access to fresh seasonal foods, digital platforms can be an important complement to autonomous learning. After all, online resources such as tutorials and instructional videos can be accessed anytime and anywhere, providing a convenient and flexible way to learn about new products or recipes, thus improving food literacy. Moreover, if digital food platforms offer limited opportunities for face-to-face interactions, the exchange between producers and consumers via online platforms appears to foster a greater understanding of the origins and seasonality of food as well as of the processes involved in its production. ('De-McDonalisation' and intellectual reskilling).

Yet our analysis revealed that informal knowledge sharing and 'learning by doing' in physical real-life contexts are crucial for facilitating the (re) acquisition of knowledge around seminal aspects of quality related to food, such as an increased consideration of producers' labour (reskilling in processual systems) and an enhanced awareness about how food choices and the food system impact society (civic reskilling) and the broader sphere of social life (social reskilling). These types of reskilling seem, in fact, to require a deeper level of emotional connection, non-verbal communication, and the ability to build trust and empathy, which appear to be difficult to achieve in virtual settings. In this regard, food procurement channels that enable peer-to-peer encounters and extend beyond merely providing 'good food' appear to be better placed to help the consumer regain these competencies and foster food consumption that goes beyond purchasing, handling, and sharing food within households (Dal Gobbo et al., 2022). Thus, to cultivate a complete understanding of the complex elements of quality related to food, it is important to find a balance between the convenience and accessibility of digital platforms and the value of face-to-face interactions in grassroots AFNs. In this sense, grassroots digital food platforms seem to provide a good example of this possibility, although they currently represent only a small proportion of the available options for consumers.

A more sustainable food system requires a more informed and knowledgeable citizen-consumer, able to resist the lure of packaged-food cuisine and ready-made meals. If purposefully designed, new 'online' alternative food networks, as a socio-technical innovation in the local food economy, could provide an opportunity to enhance citizens' ability to regain control over their food choices and contribute towards a more sustainable society. Of course, in order to be oriented towards the common good, their transformative capacity should be integrated within broader ideas of food system transformation and food justice, such as those that are stimulating the spread of local food policy councils, where food consumption is seen as just one of the vehicles for building and sustaining alternative forms of sociality, fairer economies, and healthier ecosystems.

## References

Bos, E., & Owen, L. (2016). Virtual reconnection: The online spaces of alternative food networks in England. *J. Rural Stud., 45*, 1–14. https://doi.org/10.1016/j.jrurstud.2016.02.016

Brunori, G., Rossi, A., & Guidi, F. (2012). On the new social relations around and beyond food. Analysing consumers role and action in Gruppi di Acquisto Solidale (solidarity purchasing groups). *Sociologia Ruralis, 52*(1), 1–30. https://doi.org/10.1111/j.1467-9523.2011.00552.x

Carrigan, M., & Attalla, A. (2001). The myth of the ethical consumer—Do ethics matter in purchase behaviour? *Journal of Consumer Marketing*. MCB UP Ltd, 560–578. https://doi.org/10.1108/07363760110410263

Colatruglio, S., & Slater, J. (2014). Food literacy: Bbridging the gap between food, nutrition and well-being. In F. Deer, T. Falkenberg, B. McMillan, & L. Sims (Eds.), *Sustainable well-being: Concepts, issues, and educational practices* (pp. 37–55). ESWB Press.

Cox, R., Holloway, L., Venn, L., Dowler, L., Ricketts, J., Kneafsey, H. M., & Tuomainen, H. (2008). Common ground? Motivations for participation in a community-supported agriculture scheme. *Local Environment, 13*(3), 203–218. https://doi.org/10.1080/13549830701669153

Cui, Y. (2014). Examining farmers markets' usage of social media: An investigation of a farmers market Facebook page. *Journal of Agriculture, Food Systems, and Community Development, 5*, 87–103. https://doi.org/10.5304/jafscd.2014.051.008

Dal Gobbo, A., Forno, F., & Magnani, N. (2022). Making "good food" more practicable? The reconfiguration of alternative food provisioning in the online world. *Sustainable Production and Consumption, 29*, 862–871. https://doi.org/10.1016/j.spc.2021.07.023

De Bernardi, P., Bertello, A., & Venuti, F. (2019). Online and on-site interactions within alternative food networks: Sustainability impact of knowledge-sharing practices. *Sustainability (Switzerland), 11*(5), 1457. https://doi.org/10.3390/su11051457

Dwiartama, A., & Piatti, C. (2016). Assembling local, assembling food security. *Agriculture and Human Values, 33*(1), 153–164. https://doi.org/10.1007/s10460-015-9624-9

Etmanski, C., & Kajzer Mitchell, I. (2017). Adult learning in alternative food networks. *New Directions for Adult and Continuing Education, 2017*(153), 41–52. https://doi.org/10.1002/ace.20220

Feagan, R., & Henderson, A. (2009). Devon acres CSA: Local struggles in a global food system. *Agriculture and Human Values, 26*(3), 203–217. https://doi.org/10.1007/s10460-008-9154-9

Forno, F., Grasseni, C., & Signori, S. (2015). Italy's solidarity purchase groups as citizenship labs. In E. Huddart Kennedy, M. J. Cohen, & N. Krogman (Eds.), *Putting sustainability into practice. Applications and advances in research on sustainable consumption* (pp. 67–88). Edward Elgar Publishing.

Forssell, S., & Lankoski, L. (2015). The sustainability promise of alternative food networks: An examination through "alternative" characteristics. *Agriculture and Human Values, 32*(1), 63–75. https://doi.org/10.1007/s10460-014-9516-4

Fuentes, C., & Fuentes, M. (2022). Infrastructuring alternative markets: Enabling the exchange of local food through patchworking. *Journal of Rural Studies, 97*, 13–22. https://doi.org/10.1016/j.jrurstud.2022.05.022

Goodman, D., Du Puis, M., & Goodman, M. (2012). *Alternative food networks.* Routledge.

Grasseni, C. (2013). *Beyond alternative food networks: Italy's solidarity purchase groups.* Bloomsbury Publishing.

Jaffe, J. (Ed.). (2015). *Deskilling. The SAGE encyclopedia of food issues* (pp. 351–356). SAGE Publications, Inc.. https://doi.org/10.4135/9781483346304

Jaffe, J., & Gertler, M. (2006). Victual vicissitudes: Consumer deskilling and the (gendered) transformation of food systems. *Agriculture and Human Values, 23*, 143–162. https://doi.org/10.1007/s10460-005-6098-1

Kloppenburg, J., Hendrickson, J., & Stevenson, G. W. (1996). Coming in to the foodshed. *Agriculture and Human Values, 13*(3), 33–42. https://doi.org/10.1007/BF01538225

Lewis, T. (2020). *Digital food: From paddock to platform.* Bloomsbury Academic.

Marsden, T., & Sonnino, R. (2012). Human health and wellbeing and the sustainability of urban–regional food systems. *Current Opinion in Environmental Sustainability, 4*(4), 427–430. https://doi.org/10.1016/j.cosust.2012.09.004

Milestad, R., Westberg, L., Geber, U., & Björklund, J. (2010). Enhancing adaptive capacity in food systems: Learning at farmers markets in Sweden. *Ecology and Society, 15*(3), 29. http://www.ecologyandsociety.org/vol15/iss3/art29/

Miralles, F., Giones, F., & Gozun, B. (2017). Does direct experience matter? Examining the consequences of current entrepreneurial behavior on entrepreneurial intention. *International Entrepreneurship and Management Journal, 13*, 881–903. https://doi.org/10.1007/s11365-016-0430-7

Moore, J. W. (2015). Cheap food and bad climate: From surplus value to negative value in the capitalist world-ecology. *Critical Historical Studies, 2*(1), 1–43. https://doi.org/10.1086/681007

Patton, M. 2002. *Qualitative research and evaluation methods.* 3rd ed. Thousand Oaks, CA: Sage.

Plessz, M., & Étilé, F. (2019). Is cooking still a part of our eating practices? Analysing the decline of a practice with time-use surveys. *Cultural Sociology, 13*(1), 93–118. https://doi.org/10.1177/1749975518791431

Ritzer, G. (2000). *The McDonaldization of society* (new century ed.). Pine Forge Press

Savarese, M., Chamberlain, K., & Graffgna, G. (2020). Co-creating value in sustainable and alternative food networks: The case of community supported agriculture in New Zealand. *Sustainability (Switzerland), 12*(3). https://doi.org/10.3390/su12031252

Schlosser, E. (2012). *Fast food nation: The dark side of the All-American meal.* Mariner Books.

Schneider, T., Eli, K., Dolan, C., & Ulijaszek, S. (Eds.). (2019). *Digital food activism.* Routledge.

Shapiro, L. (1986). *Perfection salad.* Farrar, Straus, & Giroux.

Signori, S., & Forno, F. (2019). Consumer groups as grassroots social innovation niches. *British Food Journal, 121*(3), 803–814. https://doi.org/10.1108/BFJ-08-2018-0523

Slater, K., & Robinson, J. (2020). Social learning and transdisciplinary co-production: A social practice approach. *Sustainability (Switzerland), 12*(18). https://doi.org/10.3390/su12187511

Stuiver, M. (2006). Highlighting the retro side of innovation and its potential for regime change in agriculture. *Research in Rural Sociology and Development, 12*, 147–173. https://doi.org/10.1016/S1057-1922(06)12007-7

Sundkvist, Å., Milestad, R., & Jansson, A. M. (2005). On the importance of tightening feedback loops for sustainable development of food systems. *Food Policy, 30*(2), 224–239. https://doi.org/10.1016/j.foodpol.2005.02.003

Thorsøe, M., & Kjeldsen, C. (2016). The constitution of trust: Function, configuration and generation of trust in alternative food networks. *Sociologia Ruralis, 56*(2), 157–175. https://doi.org/10.1111/soru.12082

Veen, E. J., & DAmico, S. (2019) Practices of food provisioning in alternative food networks: How different practitioners engage in different practices, depending on their emotional energy. *International Journal of Sociology of Agriculture and Food, 25*(1), 1–20. Retrieved January 5, 2021, from http://www.ijsaf.org/index.php/ijsaf/article/download/20/43

Warde, A., Cheng, S.-L., Olsen, W., et al. (2007). Changes in the practice of eating. *Acta Sociologica, 50*(4), 363–385. https://doi.org/10.1177/0001699307083978

Wenger, E. (1998). Communities of practice: Learning, meaning, and identity. *Cambridge University Press.* https://doi.org/10.1017/CBO9780511803932

Zoll, F., Specht, K., Opitz, I., Siebert, R., Piorr, A., & Zasada, I. (2018). Individual choice or collective action? Exploring consumer motives for participating in alternative food networks. *International Journal of Consumer Studies, 42*(1), 101–110. https://doi.org/10.1111/ijcs.12405

# Food, Health, and Sustainability: Choice, Care, Alternatives

*Alice Dal Gobbo*

## 1 Introduction and Context

Food—in all its cultural, material, discursive, and inter-subjective aspects—is a key part of everyday life. It holds a special place among the things people regularly deal with, external from and yet organically merging with them. It roots human bodies to the earth, connects them with flowing water and the surrounding air. Food consumption is a matter of bare survival and yet is also profoundly cultural, imbued with social meanings, affects, and relations. Food consumption, if experienced with mindfulness and attention, promotes a perception of the self as relational, material, and embodied: not a bounded, self-sufficient, and sovereign individual but rather a radically open and fragile subject who affects and is affected by the world at large—human and non-human (Mol, 2008). Thus, unsurprisingly, food is one key area where issues of health and sustainability

A. Dal Gobbo (✉)
Department of Sociology and Social Research, University of Trento, Trento, Italy
e-mail: alice.dalgobbo@unitn.it



A. Dulsrud, F. Forno (eds.), *Digital Food Provisioning in Times of Multiple Crises*, Consumption and Public Life,
https://doi.org/10.1007/978-3-031-46323-5_8

intersect nowadays, bringing to the fore the intimate relationship between human and ecosystems' wellbeing. Questions regarding 'eating well' also encounter wider political ones, concerning what 'living well' means.

The digitalisation of food consumption, for instance through platforms and apps, introduces novel questions and challenges in this context. Healthy and sustainable eating are often understood as related to direct, human, embodied relations among consumers, producers, and territories. Unmediated and material knowledge seems to imply more 'authentic' and natural food experiences. On this basis, Information and Communication Technologies (ICTs) might appear at odds with consumers' agendas, introducing an 'immaterial' and artificial distance from food. Nevertheless, recent examples of emerging food platforms show that innovative technologies might foreground a 'virtual reconnection' (Bos & Owen, 2016) along the food chain, becoming the very vehicle for access to products that consumers identify as healthy, sustainable, and ethical (Dal Gobbo et al., 2021).

The relationship between health, food, and what we today call 'sustainability' has a long history, which intersects with social movements. From Rachel Carson's *Silent Spring* (2002) to ecofeminist scholarship and environmental justice activism, this nexus has been politicised, criticising and resisting capitalist patriarchal and colonial models of food production and consumption, which regularly sacrifice territories, bodies, and entire (food) cultures to profit and the imperatives of global capital (Gaard, 2011; Shiva, 1988; Salleh, 2016). Nowadays, awareness of industrial farming's toxicity and the unhealthiness of associated lifestyles are far more widespread, accepted, and becoming mainstream also in policymaking arenas (see, for instance, the recent *Farm to Fork* strategy by the European Commission in the context of the European Green Deal (2020)). This consolidation of food, health, and sustainability as a nexus in scientific discourses, popular knowledge, public policies, and everyday practices has strongly influenced the dietary habits of many people, especially among the middle and upper classes of advanced industrial countries, raising important issues regarding the exclusionary effects of this rhetoric. Alternative forms of eating grounded in different cultures are constructed as unhealthy, and thus irresponsible and even unsustainable, reinforcing the ethnocentrism of contemporary understandings of 'proper diets' (Martínez, 2015). Also, issues of justice emerge towards nonconforming bodies such as those of fat and obese persons, as the presumed unsustainability of their diets becomes a further means of body shaming (Mackert,

2022). Finally, the very idea that the pursuit of healthy lifestyles might also favour sustainable food practices (Sassatelli, 2004; Lea, 2005; Barosh et al., 2014; Shugart, 2014; Apaolaza et al., 2018) is not necessarily always valid: superfoods are considered healthy and yet problematic ecologically (Magrach & Sanz, 2020); similarly, fish is deemed healthy, but fishing's ecological side effects are very profound (Brunner et al., 2009).

Drawing on 45 qualitative in-depth interviews carried out in Milan, Italy, between June 2019 and August 2020, this chapter investigates the health–sustainability nexus from the perspective of everyday life, looking at how these concerns assemble in food consumption practices. Consumption is a key part of sustainable transitions, since it can steer (although certainly not determine) production towards more sustainable practices, supporting virtuous economies; it also has direct impacts on food sustainability because it involves planning, waste, and choices regarding what and how to cook, share, preserve and finally dispose of. My specific focus will be on consumers who shop through provisioning channels that self-define as 'sustainable', like organic supermarkets, platforms for the sale of local and fresh produce, and alternative food networks (AFNs). I understand these channels as 'innovative', in as far as they propose a model of distribution and consumption that goes beyond hegemonic mass distribution, introducing social and/or technical innovations: digital technologies, horizontal and self-managed relations, direct contact between producers and consumers, etc. These channels play a key role in making 'good food' (healthy, quality, diverse) available to urban dwellers (Sage, 2003), yet in very different ways. Before proceeding to the analysis, I contextualise my approach within the wider debate and literature on health, food, and sustainability. Then, I briefly present methods and practical matters concerning research. In Sect. 4, divided into three subsections, I go into detail about research observations with the aid of empirical material. In the conclusion, I discuss emerging reflections and draw lines for future research, with a special focus on challenges and open questions digital platforms introduce in the context of sustainable food consumption.

## 2 Food, Health, and Sustainability in Context

Literature in the social sciences of food has pointed out the relationship between food practices and discourses of health and wellbeing (Schaefer et al., 2016). What in the US has been identified as an 'epidemic' of obesity and other noncommunicable diseases (NCDs) attributed to unhealthy eating and lifestyles has brought attention back to what had remained for some time a residual practice. Especially in a neoliberal context of welfare retrenchment, given the costs of public healthcare, promoting healthier diets and lifestyles among the population has become a key priority for policymakers (Henwood et al., 2011; Blue et al., 2016). Campaigns, official media coverage, public health policies, and shared meanings over the eating–wellbeing link have made food an increasingly important dimension of everyday self-care. The effort in this sense has been so vast and penetrating that today, in countries such as the US and Canada, health is the theme that most commonly accompanies food concerns (Shugart, 2014).

Healthy diets are defined as balanced and diversified regarding their nutrient content, regarding whether they include large quantities of fruits and vegetables and regarding whether they reduce refined carbohydrates and meat intake—while preferences are expressed for wholemeal cereals, legumes, and unprocessed ingredients (Eriksson & Machin, 2020). Normally, organic food is associated with healthiness, although provenance is also deemed crucial, with local food framed as healthier because it is fresher and seasonal (Migliore et al., 2018; Rizzo et al., 2020). Finally, taste and quality feature among the indicators of healthy food too (Feldmann & Hamm, 2015). Among the foods constructed as 'bad' are mostly ready-to-eat processed food, including frozen and sugar- and fat-rich foods and red meat. A reduction of animal products is almost universally regarded as healthy, yet vegan and vegetarian diets remain a contested issue, differently mobilised depending on specific scientific and ethical orientations (Bonnet et al., 2020). The canonical three main meals a day (breakfast, lunch, and dinner) are suggested, complemented by a couple of snacks throughout the day. At dinnertime, people are discouraged from consuming too much food too late in the evening and urged to favour lighter and more digestible ingredients (Lindsay et al., 2019). Of course, these definitions of healthy eating are not universal or neutral, although Western medical science has been powerful in constructing them as such while consolidating their normativity. The presumptions of 'scientificity' that surround contemporary discourses of healthy eating are the result of power struggles around what knowledge and practices count and around the grounds for the definition of 'good eating', whereby vernacular, situated, native cultures—and even individual idiosyncrasies—are overwritten by abstract guidelines that themselves provide a simplified account of the complexity of the scientific debate (Ristovski-Slijepcevic et al., 2010; Henwood et al., 2011; Martínez, 2015).

More recently, concerns around ecology and the sustainability of everyday eating practices are also entering the houses (and tables) of many eaters. Capitalist agribusiness, the dominant model of food production, distribution, and consumption, is detrimental not only to human health but also to that of local territories and the wider planet. As discourses of ecologic crisis, sustainability, and the Anthropocene make evident the link between food systems, human health, and planetary wellbeing, healthy eating discourses become increasingly intertwined with environmental, social, and economic sustainability concerns (Haines, 2020; Marsden & Sonnino, 2012). From this complex understanding of 'proper' diets, a wider idea of 'ethical eating' (Lang, 2010) takes hold, including reflexive forms of consumption that seek to keep together attention to bodily wellbeing, care for others in the family, sensitivity to the environment, support for local economies and small farmers, socioeconomic equity, etc. The concept of 'sustainable diet' captures these intersecting concerns, largely inheriting the principles of healthy diets, but articulating them with reference to specific territories and their food systems' affordances (Macdiarmid et al., 2012).

Evaluating how these emerging links might impact the overall sustainability of food systems, particularly with reference to emerging sociotechnical innovations for food provisioning, requires critical inquiry into how the food–health–sustainability nexus articulates. Public interventions and social marketing seek to mould or nudge people's eating attitudes, choices, and habits, emphasising behaviour, agency and sovereignty, and self-conduct (Hoek & Jones, 2011; Guthrie et al., 2015). The market is seen as the space where 'consumer-citizens' (Jacobsen & Dulsrud, 2007; Sassatelli, 2015) have the most power to choose and consumption acquires a moral dimension: choosing in the 'right' way means being responsible, not burdening the community at large and the environment (Beagan et al., 2015; Blue et al., 2016). Expert knowledge plays a key part in this, with the science of nutrition and environmental science suggesting appropriate courses of action (Neuman, 2019; Madden & Chamberlain, 2010). Yet, the simplistic and uncontroversial translation of scientific knowledge favours technocratic management of individual (and planetary) health (Henwood et al., 2011).

The mainstream constructions of sustainable and healthy eating as an individual responsibility largely overlook (and indeed obscure) the differential access to the food market, and in particular to that concerned with 'good food'. Especially in urban areas where living costs are high and the organisation of space and time is dispersive, food geographies exclude a great part of the population from ethical provisioning and eating (Vonthron et al., 2020; Halkier, 2009; Spaargaren et al., 2012). As such, 'good eating' remains the privilege of the upper and middle classes, which might use it as a means of class (and racial) 'boundary marking' (Beagan et al., 2015; Shugart, 2014; Backett-Milburn et al., 2010). Already disadvantaged subjects are blamed for unhealthy and unsustainable eating practices and women disproportionately become the bearers of the responsibility for their and their families' wellbeing and fitness (Evans, 2011; MacKendrick, 2014), reinforcing patriarchal power distribution in the family and society at large (Madden & Chamberlain, 2010; O'Neill & Silver, 2017). Meanwhile, attention is diverted from market, state, and corporate responsibilities.

In addition, such an oversimplified model of 'proper eating' excludes key dimensions of food practices, such as personal enjoyment, conviviality, sociocultural meanings, propensities, habits, and life contexts (Feldmann & Hamm, 2015; Shugart, 2014). It is unable to resonate with the concreteness of daily consumption (Warde, 2015). For instance, food consumption in the family is seldom a simple matter of putting the 'right' foods on the table: it involves mediation among different needs and desires, considering tastes, dietary needs, and the will to care for significant others. Consumer-citizens often feel overburdened with a responsibility of choice though they largely do not have choice. They perceive a *lack* of control over purchases, denouncing the utter difficulty and impossibility of choice, even as they strive to comply with healthy and sustainable lifestyle standards. Information is often ambiguous, uncertain, and unreliable, leading to frustration and anxiety, with a poorly regulated and little transparent use of labelling (Oleschuk, 2020). The 'technologisation of communication' manipulates and commodifies values of health, ethical worker's treatment, and environmental sustainability (Eriksson & Machin, 2020).

Apart from the responsibility that consumer-citizens perceive, other attitudes seem to foreground sustainable and healthy food consumption. As Brettin et al. (this volume) suggest, noninstrumental and other-oriented attitudes of care are crucial—albeit not always framed explicitly in terms of ethical commitment (Ančić et al., 2019; Smith et al., 2015). For instance, at childbirth, mothers and fathers start to question food practices because they feel more other-oriented and responsible for the child (Hall, 2011). Similarly, preoccupations over the environmental conditions of one's children's future are part of the everyday ethical evaluation of actions, which open the present to extended temporalities and lives-to-come (Groves, 2009). On another level, grassroots social movements and initiatives (e.g., Alternative Food Networks and the permaculture movement) collectively practice everyday more-than-human care, building infrastructures for diverse, healthy, and sustainable food systems (Goodman et al., 2013). Based on nonreductionist ideas of health, wellbeing, and sustainability, dynamic interrelations and material exchanges frame individual wellbeing as directly dependent on the planet's and other people's wellbeing (Haines, 2020; Centemeri, 2018). Through collective participation, knowledge sharing, and common practices, these experiences politicise the link between the body in its situated, singular, existence and the sociopolitical, economic, and ecologic environment in which it is located (Mallory, 2013).

The advent of digital platforms for food shopping introduces ambivalent elements into this complex picture. The omnipresence and supposed ease of use of and access to ICTs have brought optimism for their capacity to widen the reach of potential consumers, making it possible, for instance, to purchase ethical food without long-distance travelling to dedicated shops, with the convenience of flexibly organising time (Samsioe & Fuentes, 2022; Schneider & Eli, 2022). There is, indeed, evidence that platforms increase access to 'good food' for people in urban centres (Dal Gobbo et al., 2021). Platforms, by their very constitution that favours horizontal relations among actors, also seem promising in enabling the kind of reconnections between producers and consumers that AFNs have sought to establish for decades; they become the very instruments of activism (Bos & Owen, 2016). Reestablishing contact might increase transparency and consumers' trust in the products they buy, as well as increasing their motivation to buy sustainable foods. However, these technologies might be exclusionary due to IT (il)literacy and access, functioning (Fuentes, 2019), and price. Furthermore, digital instruments might make food shopping feel abstract and detached, alienating consumers even more from the food they bring to their tables. It is important, then, to evaluate how platform use 'encounters' everyday food logics on their very ground, without assuming an abstract and universal model of 'citizen-consumer' or technological development, but rather empirically assessing the concrete relationalities that become established around sociotechnical innovations (Southerton & Fuentes, 2022).

The following discussion focuses on how platform shopping reframes healthy and sustainable everyday food practices. Annemarie Mol's (2008; but see also Miller, 1998) conceptual distinction between the 'logic of care' and the 'logic of choice', which she saw deployed in healthcare institutions, will guide analysis and help to illuminate the dynamics informing platform use. A logic of choice informs neoliberal politics of health, where individuals are deemed responsible for their bodily conditions, shifting 'the weight of everything that goes wrong onto the shoulders of the patient-chooser' (Mol, 2008, p. xi). To the contrary, a logic of care informs healthy and sustainable food practices as subjects position themselves in relation with the world, aware that (good) living depends on collective and more-than-human wellbeing. The two concepts were chosen to guide analysis because they reflect emerging patterns in the data, although of course they are heuristic tools that do not exhaust the complexity of the data. Also, they are not mutually exclusive, because the two logics coexist throughout the data set and even in the same interview.

## 3 Empirical Material and Context

The present chapter is based on 45 qualitative interviews carried out in Milan between June 2019 and August 2020. Milan was chosen as the locale for the study because, at the time, it was the Italian city where digital platforms were used for food shopping the most. Interviewees were consumers who shop for food through innovative channels that self-define as 'sustainable'. The shopping channels selected were five, covering both online and offline, business and grassroots initiatives. As an 'analogue' business, NaturaSì shops were selected, a big nationwide distributor specialised in organic products. Online businesses included the platforms Cortilia and Food Assembly (FA), both very popular in Milan. Offline grassroots initiatives were GAS (solidarity purchasing groups): since more than 30 years, these groups allow consumers to connect directly to selected producers to purchase convenient food that also responds to ethical and sustainability standards. The grassroots platform was Buonmercato, which was born more than 10 years ago directly out of the GAS experience. Due to the COVID-19 pandemic, the latter half of the interviews were conducted online. This contingency had dual effects on the interviews themselves. In part, the personal contact and the immersive, embodied, engagement with the spaces and matters of everyday food were lost. But digital interviews made date and time arrangements easier, while not sacrificing much of the access to households, because interviewees seemed almost more at ease in disclosing their life spaces via the intermediation of the camera.

Interviewees included a fair degree of variation. They ranged from 23 to 67 years old, with the majority falling within the 40–60 age band: 29 were women, 3 were interviewed as couples, and the rest were men. Although it might be argued that all of them could be attributable to the middle classes, their socioeconomic status was highly different, spanning from university students and low-skill clerk positions to researchers, professionals, and small entrepreneurs; some interviewees were retired and two were housewives. Educational levels were also rather differentiated, although most interviewees had at least a university degree. Most people interviewed for NaturaSì, FA, and Cortilia resided in the central or semi-central areas of Milan, whereas some of the GAS and Buonmercato consumers resided on the outskirts of Milan and even outside of it (Buonmercato is indeed based in Corsico, an independent council). The family status of interviewees ranged from independently living singles to students living with their parents, and from families with young children to couples whose children had left home.

The semi-structured interview schedule did not include any questions around health and wellbeing. Yet, some questions were particularly potent in triggering these discourses, making participants' concerns evident. Significantly, many raised the health issue right from the start, while they were presenting themselves: as they introduced the reasons behind their (careful) food choices, health emerged as one key element. These concerns surfaced again as people were asked about what they liked most about the food channel at issue, compared to other ones. Questions around the 'proper' meal (Murcott, 1982) made food moralities and discourses around 'good' ways of eating emerge. Finally, as people articulated the changes they desired to implement to their diet, health came up once again as a significant concern. Although health and wellbeing were shared issues throughout the data set, they were more explicitly framed as a priority by NaturaSì consumers, whereas digital and GAS consumers seemed to intersect them, to varying degrees, with other ethical and convivial concerns that made them important, but not as determinant, in food practices.

## 4 Analysis: Food, Health, and Sustainability in the Making

Health concerns, despite not being an explicit subject of the interviews, are pervasive throughout the data set, deeply influencing food practices and familial relations. Although environmental sustainability seems to be a less pressing preoccupation, interviewees make recurrent links between human and non-human health: that 'everything is interconnected' is something they seem to almost take for granted. A broad understanding of 'good food' is mobilised, where healthiness goes hand in hand with respect for the environment and other people (see Sage, 2003):

> Yes [eating organic] is a health matter mainly but I think that it's all linked really that what is modified it means that it is altered and generally speaking altering a natural process is a forcing that I don't like I mean I don't like it even if I do it on myself. (AR, female, 39 years old, NaturaSì consumer)

In respondents' perception, the food in mainstream channels is both unhealthy *and* socioenvironmentally damaging, based on exploitation of human and non-human beings. The main reason for virtually all our participants to approach 'alternative' provisioning channels is a will to feed themselves and their family food that tastes good, but that also recalls ideas of naturalness, artisanal production, small-scale and humane farming, and even spiritual dimensions. In particular, food platforms are attractive because they are perceived as more practicable and accessible (Dal Gobbo et al., 2021). Yet, as said, these preoccupations 'assemble' in very different ways, with different implications for sustainability.

### *The Logic of Choice: 'Fitness' as Morality*

> As a citizen I have an ideal maybe because I am the daughter of a physician and he taught me the importance of prevention … and this touches the theme of food as well my mother introduced me to macrobiotics that at the time was really disgusting but the ideal of organic food has remained with huge efforts because I love food as you can see I am not one of those people of little appetite thin slender but anyway I try to maintain my body in decent shape above all I would like it to be performing because a people like us we are old it is nice that it keeps fine at sixty-five, seventy-five … my uncles from Fiume they are 85 and walk without any problem no Alzheimer no mental retardation senile dementia nothing because these people despite not having

> this modern food culture but the old one let's say they managed to get to their age without paddies and these terrible ailments that many old people use going around with walking sticks and on top of this they are such a burden on our payment balance because the State must take care of them I would like to be one of those whom the state does not need to take care of and nutrition is crucial because it can make you ill but it can also heal you if you develop your knowledge. (CB, female, 66 years old, NaturaSì consumer)

It is worth quoting CB's extract at length, because it encapsulates many features that characterise the logic of choice in health and nutrition. First, the concern for a *performing body*, which food is crucial to cultivate (Sassatelli, 2000). Ageing processes imply fragility and a growing dependence on proximate others and in CB's words themselves, welfare-state provisioning. In line with neoliberal moralities, CB reiterates the idea that a good citizen should not become a burden for the collective. Diet is key for achieving this: preserving an autonomy that is understood in a strongly individualistic sense. 'Decency' equals being an able body who can manage everyday life without entering relations of dependency with others, avoiding using devices that signal fragility, such as diapers or walking sticks. CB thus tries to maintain a body that is *fit* to meet current societal demands, as is also evident from her repeated references to aesthetic standards, weight, and appearance.

This example shows how bodies and subjectivities are traversed by a 'biopower' (Rabinow & Rose, 2006) that encourages self-control to govern population, respond to the capitalist need for active labour power, and alleviate state responsibility caring for citizens (Curtis, 2002; Rose, 2001). Intensification and potentiation of bodies through diet become key parts of common understandings of food. As also other participants testify, dominant economic rationalities are integrated in everyday decision-making (Pellandini-Simányi, 2021), and healthy eating is instrumentally understood as an 'investment' in the future, which will allow reducing spending for medications and cures. Similarly, several healthy eaters never buy intensively farmed meat or fish and sometimes they are vegan, pushed not so much by a will to end animal suffering but rather to avoid that the animals' pain, embodied in their 'flesh and nerves', might have deleterious effects on their health.

The morality of healthy eating as guided by a logic of choice often implies sacrifice. As CB herself admits, it is 'hard' to always eat in the 'right' way, but she builds a whole discipline and self-restraint in a sort of precarious and dynamic search for balance between food enjoyment and 'proper' diet. All this involves expensive and demanding practices such as eating her lunch every day in a restaurant owned by a healthy food guru, buying organic products from NaturaSì shops, a labour of food selection that she has carried out, as she likes to say, 'throughout the centuries'. My afternoon tea with GO (female, 29 years old), again a NaturaSì consumer, confirms this, as she laments that the organic biscuits we are eating are not very tasty: '[T]hey do not use stuff like preservatives so it is obvious that they have simpler tastes than industrial ones and you are used to those … you taste the difference a bit but you make an effort because the primary objective is to be well'. Although this is commonly less the case when talking about fresh products such as fruits and vegetables, 'being well' seems to involve a sacrifice in taste that is made for a higher good: health.

Knowledge, and above all expert knowledge, plays a key role in shaping the moralities of healthy eating and the very practices that correspond to it. Physicians' and dietary scientists' recommendations are seen as recipes not only for health but also for a generalised wellbeing, sometimes charged with spiritual overtones; even when interviewees do not follow dietary gurus, more generalist 'science' and 'information' play a key role in determining food practices. They shape everyday habits and rhythms, the choice of where and what food to buy, how to cook it, when to eat it, how and with whom to share it, etc. Expensive retailers such as NaturaSì are often the preferred venues in shopping-for-health. Interviewees report going there more than once a week to always have fresh foods in their pantry; they avoid ready-made foods and experiment with often expensive superfoods for improved wellbeing. This is largely recognised as a very demanding and sometimes frustrating labour, and yet an important part of being 'good' consumer-citizens. Furthermore, these shops are not evenly distributed throughout the city, so they are partially inaccessible to many consumers. Food platforms seem to ease healthy-sustainable food availability, but again they involve complex arrangements to receive or pick up the food, and this complexity sometimes discourages purchasing; so many admit their online shopping is not enough for their week and they must therefore do supermarket shopping as a backup.

These healthy eating practices largely correspond to sustainable diet requirements, reducing the environmental impact of the food, its waste, and packaging. Still, choices in food consumption are often individualistic and instrumental. Trade-offs are constantly made and negotiated in everyday choices, and subjects tend to devalue and disregard all those consumption aspects that they do not perceive to have direct impacts on individual wellbeing. For instance, paying more for organic food is rebalanced with buying from discount supermarkets cheap and nonecological detergents and other commodities, evidently perceived as less central to individual wellbeing, as AC (female, 61, FA consumer) recounts in her interview. Furthermore, organic, sustainable, and healthy eating tends to be reserved to deserving bodies. AR (female, 39, NaturaSì consumer), for instance, is very careful in food shopping for herself, her husband, and their child, but she buys cheap food for the old uncle she looks after, because she feels that his old, decaying, body does not deserve such care any longer.

The deployment of individualistic and instrumental food consumption choices makes invisible possible collective pathways to tackle health and sustainability challenges. Participants like AR or GO do link air, soil, and water pollution to poor health, conscious that in highly urbanised environments many polluting activities are carried out in relation to mobility, industry, waste management, etc. The latter are likely to be an important cause of diseases and other adverse bodily conditions. But this situation is talked about as unchangeable and out of their control; therefore, the common argument goes, they 'at least' try to care about something they feel a hold on: food consumption. Healthy eating thus becomes a form of 'immunisation' (Esposito, 2004) response of self-preservation, as well as a self-enhancing practice through supplements and costly superfoods.

Although part of this way of dealing with health issues arises from subjectivation processes that involve internalisation of neoliberal logics and discourses, the sociomaterial assemblages of food consumption also shape affordances and choices. At times, the individualised pursuit of healthy eating is just the second-best option, and yet the only available one. This limitation comes through in a particularly strong way in the story of SP (female, 29 years old), who has used the platform Cortilia for a few months. Coming from a rural village in the northeast of Italy, she used to buy local fresh food directly from producers and to grow fruits and make fruit preserves herself. This style of provisioning she identifies as healthy and appreciable, something that she has 'learned' from her family practices, in particular from her mum's habit of going to nearby farmers, mindful of the nutritional aspect of their meals. Here in the city, she finds this style of provisioning impossible: she can access only supermarkets and a medium-quality greengrocer. A platform like Cortilia provides a means of accessing food of better quality, 'not even comparable' with the rest—albeit at a price that is significantly higher too.

SP mentions healthiness and fitness as part of her motivations to buy via the platform. But throughout her narrative, it emerges that sharing and appreciating good food (also in the company of her boyfriend, who actively participates in the choices and was indeed the initiator of the commitment to Cortilia) are also, and maybe foremost, matters of cultivating better socioecological relations. She wants to become vegetarian and reads about climate justice issues; she is mindful about farming practices, labour, and the environmental impact of packaging. In these respects, she would value a provisioning style that allows her to build relationships of trust with farmers, support local small-scale economies, control non-human animals' living conditions, and appreciate the intensity of food pleasures also through the skills involved in handling food (cooking, preserving, etc.). Cortilia seems incapable of enabling this wider idea of proper eating. As I ask her what she would change in her everyday food practices, she replies:

> Yes I would like in this sense also the stuff about waste and packaging I see it as unfeasible. Cortilia is an opportunity that is given to me where anyway I use an app, anyway things get to me with the packaging, and the fact that it is possible to have the food delivered on Sunday mornings as well. … In some ways, I feel sorry about all this I don't know I have strange feelings with respect to this thing it is contaminated anyway this relation that I have with the woman that plants the apples next to my house and she makes the apple juice that at home is never missing. … Claudia, she does not exist, even if the stuff arrives from a farm that is probably like this.

In this extract, she voices preoccupations that are about the environment (packaging), social justice (conditions of labour, delivering rhythms), and collective wellbeing: the human relations that are part of the very fabric of daily living. Cortilia only provides an online description of farms and its delivery programme puts a distance between her and the concrete work and life of producers; therefore, it disenables direct contact; in some cases, and notably with respect to logistics, she even sees it as exacerbating the problems of an exploitative economy. SP is 'given' this opportunity of choice by the market, but as a ready-made and very partial solution. As GA (male, 32, Cortilia consumer) confirms, the platform surely connects, but it also divides: '[T]hey can do as nice a description as they want about the producer, [but] you don't know what's happening in the field'.

Such a situation is not due to the platform itself, but to the wider spatial and time organisation of urban life, which disables SP in her strivings towards an extended practice of socioecological care through food consumption. This is especially true in relation to her job, which she finds exhausting: a work that she has gained with great dedication and effort and about which she is 'happy, but it's a great mess and it conditions a lot my personal life … because I have little time. As a person, I am very interested in all that is not work and with this job I cannot dedicate to anything else: physical activity, personal relations, food, clothes.' Before coming to Milan, she would work, but 'I would arrive home at 5 pm and I would cook so many things, in the oven, in the pan. … I would cook really a lot and I liked it … now cooking comes down to the very aim of putting a meal together, there's no longer pleasure in putting myself at the hob.' Cortilia responds to a quite specific need or desire, buying quality food, but the platform does not represent a challenge to the material organisation of everyday life. To the contrary, platforms seem to accommodate the hectic rhythms of urban consumption, at the risk of fostering an individualised and uncritical pursuit of healthy eating compatible within existing and unsustainable orders of production, distribution, and consumption. Overall, then, the digitalisation of consumption might promote and enable individualised practices of (self-)care that remain the domain of a few privileged subjects, creating market niches that comfortably sit alongside dominant ones (Guthman, 2008)—leaving intact the very premises of a way of life that structurally promotes unhealthy, (self-)exploitative economies and forms of labour, impersonal social intercourses, and detached and abstract relations with the organic matter that sustains life and with the wider environment.

### *The Logic of Care: Nurturing*

As SP's case introduces, the logic of choice gives just a partial account of healthy eating practices, which are often guided by different concerns. Above all, they involve preoccupations for the wellbeing of proximate others, such as family members, and for social and environmental justice more generally (Miller, 1998; Cairns et al., 2013). In this case, health is constructed relationally, as something to be preserved and cultivated through shared practices with the aim of being well together. A 'logic of care' informs these healthy food assemblages, which becomes particularly evident when experiences of vulnerability and fragility rupture everyday routines. Illnesses and other adverse bodily conditions—experienced directly or by proximate others—are triggers of change. Disparate events, from having candida to suffering overweight-related diseases, through to diabetes and high blood pressure, are often the turning points at which current food practices are perceived as subjectively unsustainable but also symptoms of a wider unsustainability. These events bring sudden changes to everyday food practices, and they introduce concerns for the wellbeing of non-human animals, environmental health, and social justice.

Interviewees start using innovative platforms to search for modes of provisioning that can satisfy these concerns. For instance, CZ (female, 43, GAS consumer) tells me that the disease of a close young relative pushed her and her family towards deeper care in food choices, and on that occasion she approached her GAS. But as articulated, according to a logic of care, healthy food is not (mainly) a way of guarding oneself from contamination. Rather, it is a dynamic search for being well in one's body and its limits, together with other bodies (family members, friends, extended more-than-human communities). The examples of interviewees who live with diabetes, PV (male, 70, GAS consumer) and SS (51, male, GAS consumer), are particularly telling in this respect. The purchase (and self-production) of food is, for them, an experience that enables a better 'quality of life' because it introduces them to natural, simple, and yet wholesome foods that they profoundly enjoy, especially because they also involve sharing and collective practices. For them, eating in a healthy way is not a matter of self-enhancement through sacrifice, but rather the convivial pursuit of enhanced pleasures, in a fine balance with their bodies' fragilities, dependences, and affordances. Also, this pursuit is inextricably interweaved with concerns for environmental sustainability and social equity that they cultivate through the GAS's collective practices. As SS says: '[W]hat is the point in me saving myself alone?'

Another case in which a logic of care shapes everyday food assemblages regards experiences of parenting. The arrival of a child sensitises parents, especially mothers, to diet and the health effects of what they eat. AR (female, 39 years old, NaturaSì consumer), for instance, admits that she was committed to junk food before becoming a mother. Ever since she became pregnant, care for her child's health has become a key priority: conscious of the complete interdependence with her child's body, and of the latter's fragility, she became very attentive to food and diet. For IP (female, 43, Food Assembly consumer), instead, a concern around children's dietary education has been transformative. She started buying food from Food Assembly several months ago, when she went to a nutritionist. The main point was not about losing weight, but rather finding someone who might 'support us in taking care of our nutrition'.

IP is married and has a 15-year-old son and a 13-year-old daughter. Throughout this time, she has not been very attentive to their diet, also because the husband—a better cook than her—was completely in charge of cooking (although the responsibility for shopping is shared). Of late, she realised that the two children were not eating well, especially when alone, and she attributes this to the poor nutritional education that they have passed on. Contextually, the husband has gained weight. Knowing about family nutrition programmes that were available in a close by clinic, she decided to enrol. These programmes were fully booked, 'so in the end I went myself to the nutritionist' and embarked in an individual programme. But, as she admits, 'I ask more for my children than for myself… it's not a family program but I interpret it like this'. Apart from providing information around the proper way to balance their meals, the nutritionist daily sends her via WhatsApp recipes and cooking suggestions.

Since the advent of these changes, the food balances in the family 'are being redefined'. On the one hand, they try to follow the suggestions, on the other, the husband 'continues to cook the things that he has always done', and she integrates with healthy recipes, which she likes a lot but are not always appreciated by the rest of the family. She also prepares things like mixed cereals for the morning, or healthy snacks, for which she needs to make a careful search of ingredients in specialised shops, notably NaturaSì. The new tastes are not always in line with the children's, so she also needs to carefully deal with these novelties because 'the relation is put to the test': she holds and supports them as they become frustrated by what they perceive as sacrifices, she carefully assesses and 'negotiate' when and if it is possible to propose to them unknown tastes, she evaluates the degree of appreciation of her experimentations. Differently from the logic of choice, expert knowledge here (as in similar cases) does not seem to function in the form of eating dogmas—although it must be evidenced that it continues to hold biopolitical force, especially on feminised subjects. Guided by the priority of living well rather than of fitness and perfection, IP flexibly adapts knowledge and information to this overarching life goal.

In the context of these changes, IP also introduced online shopping via Food Assembly. In this way, she can have fresh, organic, and seasonal food to compliment the overall family eating transition. She orders and buys online her grocery and goes at a fixed day to pick it up from a collection point where, at least in theory, the producers themselves bring the food—something that should increase the degree of (re)connection with the realities of production. Doing the shopping is quite undemanding, but online purchase does imply more 'labour' (Fuentes & Samsioe, 2021) as she needs to spend a lot of time cleaning and washing the food: it certainly is a healthier choice, but it brings a lot of soil and worms in an urban and unequipped kitchen! As I ask her why she chose this platform, she admits that this is a 'cyclical thing'. Throughout her life, she has thought about changing provisioning, for instance by taking part to a GAS, yet she also found that it was 'too demanding' in organisational terms and dropped the idea. About FA she liked the fact that someone else is taking care of controlling that producers are local and trustworthy, taking on themselves organisational matters. She admits to 'not have such a strong ethical commitment to face organisational troubles for an ethical matter. If I find a solution that holds together a price that is possibly a little higher but not excessive and a good quality and control over the supply chain that's okay.'

This is what she finds in the platform, which evidently enables her healthy-sustainable eating transition, plus enhanced food pleasures. In fact, while for the children the dietary change seems a sacrifice, for IP, like for the other participants mentioned above, a healthy diet is also an enjoyment: eating wholemeal natural ingredients, largely plant-based would be her ideal diet indeed. Still, this case also introduces the risks of articulating healthy eating via a logic of care. As IPs and other cases throughout the data set suggest, established gender roles continue to shape the dietary relationalities of families. Women disproportionately take on themselves the responsibility of information seeking, practical labour, and emotional work, so that, as MacKendrick (2014) would say, healthy eating implies 'more work for mother'. For the husband, cooking remains a leisure and pleasurable activity where he can express his ability as a cook, while responsibility for his own and the children's health is taken on herself by the wife. He continues to carry on previous practices in the face of needed changes, which IP puts into practice, together with the mundane commitments of getting and handling the food from FA.

The experience of platform provisioning does not in any way put into question these power balances, and this raises issues as to the capacity of online provisioning services to promote transitions to a 'just sustainability' that is not simply environmentally sound but also fair for workers and consumers. Indeed, understood in terms of wider equity and relations,

IP's commitment to health and sustainability stops at the level of her family practices. She is happy that caring for her diet goes hand in hand with caring for the planet, but she also recognises that 'these are things that you can be concerned about only if you can afford it, so the limit of all this is economic but also organisational'. Buying through the platform is a privilege, and this she acknowledges, but does not contest in practice. So, although platforms do seem to give access to the means for pursuing an extended idea of health, analysing the concrete dynamics foregrounding online food provisioning shows contradictions and inconsistencies that have to do with the very specific integration of 'sustainable' food platforms within the capitalist market, organisation of everyday life, unequal distribution of resources, and everyday reproductive labour.

### *The Care for Health as Collective and More-Than-Human Practice*

Collective and self-organised provisioning practices introduce new elements to the articulation of healthy-sustainable food assemblages, as the example of the Buonmercato platform suggests. Born from the GAS movement in Milan, it explicitly integrates the commitments to solidarity, justice, and social responsibility that are part of its repertoire. As the chapter by Brettin and colleagues in this volume also acknowledges, many Buonmercato interviewees mention health concerns as part of their choice to join the association and the platform. But differently from most other platform consumers, health preoccupations are largely overridden by those over building a better food system, made of relations of support and trust among consumers, between producers and consumers, with the wider community. As VC (45, female, Buonmercato consumer) says:

> Compared to the others we preferred Buonmercato, for sure, because we tried it and for the philosophy of making purchases together, going and meeting producers, helping producers in difficulty by buying from them, buying together avoiding waste. Cortilia is a fantastic network, but it does not have this vision, it is like a supermarket.

The difference between FA and Cortilia, on the one hand, and Buonmercato, on the other, is not only in the vision but also in the concrete provisioning practices, because the latter involve more practical and time commitment from the associates. The platform is a useful logistical tool for organising orders and deliveries, but when the collection days come, some people volunteer to help with the distribution while others who go there to pick up their orders often wait some time and have the occasion to chat among themselves. Furthermore, as an association, there are regular meetings; group trips are organised to go and meet the farmers and spend some convivial time together; conferences are organised around themes of interest—including health and nutrition.

These collective practices contextualise the theme of health both within the appreciation of the historical and sociopolitical shaping of food habits and within a collective and critical debate about what 'good food' is and what living a 'good life' means. A logic of care is deployed, according to which being healthy means being well in one's body, accepting fragility and vulnerability, developing a pedagogy of diet for wellbeing. But this care is also politicised, because the ongoing discussions and knowledge exchange among the members encourage processes of subjective transformation. FB (female, 39, Buonmercato consumer), for instance, recalls how meeting the older people in the association and the GAS dramatically changed her relationship to food and health, and with them the wider sustainability of her everyday food practices. Being 'a woman of science' (FB is a nurse), she has always been interested in the dietary aspects of health, and she follows—with flexibility—the guidelines of famous nutritionists and food philosophies such as macrobiotics. But after becoming part of these collective initiatives, 'for me ethics have been fundamental as a push in the choice of what I have been buying, it is also what pushed me beyond the healthy choice' and towards a rearticulation of diet that includes an extended socioecological care towards non-human animals (she is almost vegetarian), ecology, working conditions.

For FB, as for other participants, this pursuit is not individualised, but rather articulated through collective practices. Differently from the experience with other platforms, the one with Buonmercato never comes through as a lonely one, entailing distance among consumers or between consumers and producers. The fact that at least some of the people who are part of the association directly know the producers, and that it is possible to go and visit them, builds strong relations of trust (see also Brettin et al., this volume). Furthermore, the fact that the platform is not a business and therefore does not aim at profit makes the products far more accessible than FA and especially Cortilia. FB is very keen to underscore that while other people think that she is somehow posh because she only buys organic; the cost of her grocery shopping is comparable to that of the supermarket:

> Milan does not offer you much … we are surrounded by supermarkets. Obviously, the small producer is rarer to find let's say so the relation becomes a bit more difficult … on the other hand, [Milan] offers so much on all fronts: you can be sustainable and solidary in all if you want, and internet is crucial in this.

For her, platforms like Buonmercato and the use of social media as tools for the management of GAS activities are a way of *closing* the distance that the urban environment puts in between producers and consumers (Bos & Owen, 2016) and, even more importantly, to expand access to 'good food' beyond the niche of wealthy consumers who can afford shopping at organic shops or through Cortilia.

To be sure, it cannot be said that self-organised platforms like GAS or Buonmercato solve these issues, as access remains limited to committed—if not wealthy—subjects (Mallory, 2013). Still, they open the space and provide the practical-discursive tools for making health and care concerns part of collective projects that strive for a fairer, non-exclusionary, food system by practicing social and ecological relations of solidarity. LM (female, 68 years old) thus reconstructs the birth of her GAS:

> Then the period of the mad cow disease came and of GMOs and so together with other people as a response to this attack on our health I decided with them to organise the GAS the Solidarity Purchasing Group with people from the neighbourhood some long-standing friends with whom I had done initiatives in the neighbourhood like the old-date feminism struggling for the establishment of a grassroots family counselling for women yeah I had done also this kind of things I have made a lot of initiatives of this social kind.

As seen above, a perceived 'attack' on the body is the initiating event of a food transition and dietary change. The response is not, however, a closure into 'sanitised' individual healthy choices, but rather an opening to a creative community, in line with previous and wider-ranging experiences of activism. As LM's husband, VV (male, 70 years old), recalls: founding the GAS was an effort to 'go beyond critical consumption' by creating an alternative to the dominant forms of market exchange and consumption: producer–consumer direct sales; favouring organic production for sustainability; active solidarity with fragile subjects; Community Supported

Agriculture; projects for the re-introduction of biodiversity in the endangered territory of Milan's Parco Agricolo Sud; and open struggles against infrastructural projects and urbanisation. In these collective projects, concerns for personal health are the push for building the infrastructures that enable the concrete practice of collective, more-than-human health. The digitalisation of movement organising seems a tool to simplify these relations but, as also noticed by Brettin et al. (this volume), it does need in-person and engaged interactions to foreground significant and deep sustainable change.

## 5 Conclusions

In this chapter, the food–health–sustainability nexus has been analysed starting from the everyday food practices of people who shop through 'innovative' channels that self-define as sustainable. Although health was not a specific focus of the interviews, it turned out to be nearly omnipresent, pointing to the centrality of this preoccupation with respect to diet and nutrition. This can be seen as the result of a governmental effort to stir food behaviours towards healthier patterns to contain what seems an 'epidemic' of NCDs and other bodily conditions. In recent times, discourses of healthy food are intersecting with sustainable concerns, especially through the concept of 'sustainable diet' and 'ethical eating': nutrition patterns that are good for the body are constructed as being good for the earth too. As people look for healthy food, it seems that they also improve their diet's environmental, social, and economic sustainability. The ways these aspects concretely assemble in everyday life practices vary. I have pointed to three healthy eating assemblages which, despite not being exhaustive, synthesise emerging patterns in the data.

Shaped by a logic of choice, healthism becomes the new morality of food for citizen-consumers who adapt to the neoliberal governmentality of the body, nutrition, and the environment. Through market choices, they strive to be good subjects, 'fit' and 'performative' in contemporary society. Fragility and dependency are largely obscured and avoided, and proper diets are used both as a means of 'immunisation' from external, environmental, risks *and* as a way of potentiating and enhancing bodily capacities. Healthy eating is here an uncritical and depoliticised practice generally reserved to privileged consumers who do not question differential access to good food, nor, more generally, the unhealthy patterning of contemporary urban everyday life (Sikka, 2019). Sometimes, despite willingness to criticise this state of things, the material organisation of existence disenables the practice of alternatives—and in this case food platforms are the ambivalent providers of a tasty and safe, but not-so-ethical or sustainable food. In these cases, the positive environmental effects of caring for one's health are limited and do not constitute a radical response to the radical challenges of capitalist agribusiness. To the contrary, they revamp accumulation processes and contribute to discipline bodies according to dominant apparatuses of 'power/knowledge' (Foucault, 1980) and socially accepted standards, obscuring the systemic roots of unhealthy and unsustainable nutrition (Marsden & Sonnino, 2012; Barosh et al., 2014).

Everyday food consumption is also characterised by the very different 'logic of care'. Health concerns extend beyond individuals' bodies and include proximate and distant beings, and the wider environment. Health is not so much a matter of 'fitness' but rather one of 'wellbeing' and 'quality of life', not defined according to external standards but to the embodied and singular capabilities of concrete bodies. As such, it is less subject to neoliberal governmentality, although, as suggested, expert knowledge does work in a biopolitical way over (feminised) bodies. When a logic of care informs healthy eating, interviewees tend to question dominant individualism and re-frame the idea of proper eating to include concerns around socioeconomic and ecologic justice, solidarity, and fairness. Still, caring might go hand in hand with 'limited ethical commitments' and the reproduction of the existing and unequal power distribution (see Brettin et al., this volume). This introduces concerns over the effective promotion of sustainable food relationalities in the home and beyond.

In the last section of analysis, I have proposed that when a logic of care moves beyond the confines of the family, collective politics and grassroots mobilisation build the concrete platforms where a deeper sustainability can be daily practiced (Centemeri, 2018). Concerns for health, in this case, do not seek for 'immunity', but rather push subjects to collectively construct alternatives to dominant forms of production, distribution, and consumption. Opening to political questions over the systems of provisioning in contemporary societies, caring also means struggling for a radical change in systems of provision, lifestyles, and values. It is not simply a matter of stating an abstract link between individual and more-than-human health, but to investigate the concrete material conditions of poor health and unsustainability, while experimenting alternatives to it. In these movements, intersecting concerns for bodily and environmental health show dynamics that are largely in line with the militant and intellectual tradition of feminism, ecofeminism, and critical ecology (Mallory, 2013). The journey starts from endangered bodies, but these cannot be saved in isolation. Bodies root action in territories, and action itself extends to a more general, anti-systemic struggle. The logic of care, initially directed towards proximate endangered bodies, thus opens to include concerns for endangered ecosystems and, more generally, for other people and living beings, promoting forms of deeper and just sustainability (Agyeman et al., 2016).

## References

Apaolaza, V., et al. (2018). Eat organic – Feel good? The relationship between organic food consumption, health concern and subjective wellbeing. *Food Quality and Preference, 63*, 51–62. Available at: https://doi.org/10.1016/j.foodqual.2017.07.011

Agyeman, J., Schlosberg, D., Craven, L., & Matthew, C. (2016). Trends and directions in environmental justice: From inequity to everyday life, community, and just sustainabilities. *Annual Review of Environment and Resources, 41*(1), 321–340. https://doi.org/10.1146/annurev-environ-110615-090052

Ančić, B., Domazet, M., & Župarić-Iljić, D. (2019). "For my health and for my friends": Exploring motivation, sharing, environmentalism, resilience and class structure of food self-provisioning. *Geoforum, 106*, 68–77. https://doi.org/10.1016/j.geoforum.2019.07.018

Backett-Milburn, K. C., Wills, W. J., Roberts, M.-L., & Lawton, L. (2010). Food, eating and taste: Parents' perspectives on the making of the middle class teenager. *Social Science and Medicine, 71*(7), 1316–1323. https://doi.org/10.1016/j.socscimed.2010.06.021

Barosh, L., Friel, S., Engelhardt, K., & Chan, L. (2014). The cost of a healthy and sustainable diet—Who can afford it? *Australian and New Zealand Journal of Public Health, 38*(1), 7–12. https://doi.org/10.1111/1753-6405.12158

Beagan, B. L., Power, E. M., & Chapman, G. E. (2015). "Eating isn't just swallowing food": Food practices in the context of social class trajectory. *Canadian Food Studies / La Revue canadienne des études sur l'alimentation, 2*(1), 75–98. https://doi.org/10.15353/cfs-rcea.v2i1.50

Blue, S., Shove, E., Carmona, C., & Kelly, M. P. (2016). Theories of practice and public health: Understanding (un)healthy practices. *Critical Public Health, 26*(1), 36–50. https://doi.org/10.1080/09581596.2014.980396

Bonnet, C., Bouamra-Mechemache, Z., Réquillart, V., & Treich, N. (2020). Viewpoint: Regulating meat consumption to improve health, the environment and animal welfare. *Food Policy, 101847*. https://doi.org/10.1016/j.foodpol.2020.101847

Bos, E., & Owen, L. (2016). Virtual reconnection: The online spaces of alternative food networks in England. *Journal of Rural Studies, 45*, 1–14. Available at: https://doi.org/10.1016/j.jrurstud.2016.02.016

Brettin, S., Brükner, M., & Magnani, N. (this volume). A plate full of food and care. Investigating care practices and gender relations at the crossroads of sustainability, innovative practices and digital platforms. In *Digital food provisioning in times of multiple crises: How social and technological innovations shape everyday consumption practices*. Palgrave Macmillan

Brunner, E. J., Jones, P. J. S., Friel, S., & Bartley, M. (2009). Fish, human health and marine ecosystem health: Policies in collision. *International Journal of Epidemiology, 38*(1), 93–100. https://doi.org/10.1093/ije/dyn157

Cairns, K., Johnston, J., & MacKendrick, N. (2013). Feeding the 'organic child': Mothering through ethical consumption. *Journal of Consumer Culture, 13*(2), 97–118. https://doi.org/10.1177/1469540513480162

Carson, R. (2002). *Silent spring*. Houghton Mifflin.

Centemeri, L. (2018). Commons and the new environmentalism of everyday life. Alternative value practices and multispecies commoning in the permaculture movement. *Rassegna italiana di Sociologia, 59*(2), 289–314.

Curtis, B. (2002). Foucault on governmentality and population: The impossible discovery. *The Canadian Journal of Sociology / Cahiers canadiens de sociologie, 27*(4), 505–533. https://doi.org/10.2307/3341588

Dal Gobbo, A., Forno, F., & Magnani, N. (2021). Making "good food" more practic-able? The reconfiguration of alternative food provisioning in the online world. *Sustainable Production and Consumption, 29*, 862–871. https://doi.org/10.1016/j.spc.2021.07.023

Eriksson, G., & Machin, D. (2020). Discourses of "good food": The commercialization of healthy and ethical eating. *Discourse, Context & Media, 33*. https://doi.org/10.1016/j.dcm.2019.100365

Esposito, R. (2004). *Bìos. Biopolitica e filosofia*. Einaudi.

European Commission. (2020). *A farm to fork strategy for a fair, healthy and environmentally-friendly food system*.

Evans, D. (2011). Blaming the consumer—Once again: The social and material contexts of everyday food waste practices in some English households. *Critical Public Health, 21*(4), 429–440. https://doi.org/10.1080/09581596.2011.608797

Feldmann, C., & Hamm, U. (2015). Consumers' perceptions and preferences for local food: A review. *Food Quality and Preference, 40*, 152–164. https://doi.org/10.1016/j.foodqual.2014.09.014

Foucault, M. (1980). *Power/knowledge: Selected interviews and other writings 1972–1977*. Pearson Education Limited.

Fuentes, C. (2019). Smart consumers come undone: Breakdowns in the process of digital agencing. *Journal of Marketing Management, 35*(15–16), 1542–1562. https://doi.org/10.1080/0267257X.2019.1686050

Fuentes, C., & Samsioe, E. (2021). Devising food consumption: Complex households and the socio-material work of meal box schemes. *Consumption Markets & Culture, 24*(5), 492–511. https://doi.org/10.1080/10253866.2020.1810027

Gaard, G. (2011). Ecofeminism revisited: Rejecting essentialism and re-placing species in a material feminist environmentalism. *Feminist Formations, 23*(2), 26–53.

Goodman, D., DuPuis, E. M., & Goodman, M. K. (2013). *Alternative food networks: Knowledge, practice and politics*. Routledge.

Groves, C. (2009). Future ethics: Risk, care and non-reciprocal responsibility. *Journal of Global Ethics, 5*(1), 17–31. https://doi.org/10.1080/17449620902765286

Guthman, J. (2008). Neoliberalism and the making of food politics in California. *Geoforum, 39*(3), 1171–1183. https://doi.org/10.1016/j.geoforum.2006.09.002

Guthrie, J., Mancino, L., & Lin, C.-T. J. (2015). Nudging consumers toward better food choices: Policy approaches to changing food consumption behaviors. *Psychology & Marketing, 32*(5), 501–511. https://doi.org/10.1002/mar.20795

Haines, A. (2020). Good health in the Anthropocene epoch: Potential for transformative solutions. In W. K. Al-Delaimy, V. Ramanathan, & M. Sánchez Sorondo (Eds.), *Health of people, health of planet and our responsibility* (pp. 309–320). Springer International Publishing. https://doi.org/10.1007/978-3-030-31125-4_24

Halkier, B. (2009). A practice theoretical perspective on everyday dealings with environmental challenges of food consumption. *Anthropology of Food, S5*. https://doi.org/10.4000/aof.6405

Hall, S. M. (2011). Exploring the "ethical everyday": An ethnography of the ethics of family consumption. *Geoforum, 42*(6), 627–637. https://doi.org/10.1016/j.geoforum.2011.06.009

Henwood, F., Harris, R., & Spoel, P. (2011). Informing health? Negotiating the logics of choice and care in everyday practices of "healthy living". *Social Science & Medicine, 72*(12), 2026–2032. https://doi.org/10.1016/j.socscimed.2011.04.007

Hoek, J., & Jones, S. C. (2011). Regulation, public health and social marketing: A behaviour change trinity. *Journal of Social Marketing, 1*(1), 32–44. https://doi.org/10.1108/20426761111104419

Jacobsen, E., & Dulsrud, A. (2007). Will consumers save the world? The framing of political consumerism. *Journal of Agricultural and Environmental Ethics, 20*(5), 469–482. https://doi.org/10.1007/s10806-007-9043-z

Lang, T. (2010). From "value-for-money" to "values-for-money"? Ethical food and policy in Europe. *Environment and Planning A: Economy and Space, 42*(8), 1814–1832. https://doi.org/10.1068/a4258

Lea, E. (2005). Food, health, the environment and consumers' dietary choices. *Nutrition & Dietetics, 62*(1), 21–25.

Lindsay, J., Tanner, C., Leahy, D., Supski, S., Wright, J., & Maher, J. (2019). The family meals imperative and everyday family life: An analysis of children's photos and videos. *Critical Public Health, 1–13.* https://doi.org/10.1080/09581596.2019.1684443

Macdiarmid, J. I., Horgan, G. W., Loe, J., Fyfe, C., Johnstone, A., & McNeill, G. (2012). Sustainable diets for the future: Can we contribute to reducing greenhouse gas emissions by eating a healthy diet? *American Journal of Clinical Nutrition, 96*, 632–639.

MacKendrick, N. (2014). More work for mother: Chemical body burdens as a maternal responsibility. *Gender & Society, 28*(5), 705–728. https://doi.org/10.1177/0891243214529842

Mackert, N. (2022, April 1). "Globesity": On an objectionable narrative of global health. *Food, Fatness and Fitness.* Retrieved February 2, 2023, from https://foodfatnessfitness.com/2022/04/01/globesity/

Madden, H., & Chamberlain, K. (2010). Nutritional health, subjectivity and resistance: Women's accounts of dietary practices. *Health, 14*(3), 292–309. https://doi.org/10.1177/1363459309356073

Magrach, A., & Sanz, M. J. (2020). Environmental and social consequences of the increase in the demand for "superfoods" world-wide. *People and Nature, 2*(2), 267–278. https://doi.org/10.1002/pan3.10085

Mallory, C. (2013). Locating ecofeminism in encounters with food and place. *Journal of Agricultural and Environmental Ethics, 26*(1), 171–189. https://doi.org/10.1007/s10806-011-9373-8

Marsden, T., & Sonnino, R. (2012). Human health and wellbeing and the sustainability of urban–regional food systems. *Current Opinion in Environmental Sustainability, 4*(4), 427–430. https://doi.org/10.1016/j.cosust.2012.09.004

Martínez, A. D. (2015). The juxtaposition of comiendo bien and nutrition. *Food, Culture & Society, 18*(1), 131–149. https://doi.org/10.2752/175174415X14101814953800

Migliore, G., Borrello, M., Lombardi, A., & Schifani, G. (2018). Consumers' willingness to pay for natural food: Evidence from an artefactual field experiment. *Agricultural and Food Economics, 6*(1), 21. https://doi.org/10.1186/s40100-018-0117-1

Miller, D. (1998). *A theory of shopping.* Cornell University Press.

Mol, A. (2008). *The logic of care: Health and the problem of patient choice.* Routledge.

Murcott, A. (1982). On the social significance of the "cooked dinner" in South Wales. *Social Science Information, 21*(4–5), 677–696. https://doi.org/10.1177/053901882021004011

Neuman, N. (2019). On the engagement with social theory in food studies: Cultural symbols and social practices. *Food, Culture & Society, 22*(1), 78–94. https://doi.org/10.1080/15528014.2018.1547069

O'Neill, K., & Silver, D. (2017). From hungry to healthy: Simmel, self-cultivation and the transformative experience of eating for beauty. *Food, Culture & Society, 20*(1), 101–132. https://doi.org/10.1080/15528014.2016.1243765

Oleschuk, M. (2020). "In today's market, your food chooses you": News media constructions of responsibility for health through home cooking. *Social Problems, 67*(1), 1–19. https://doi.org/10.1093/socpro/spz006

Pellandini-Simányi, L. (2021). The financialization of everyday life. In C. Borch & R. Wosnitzer (Eds.), *The Routledge handbook of critical finance studies.* Routledge.

Rabinow, P., & Rose, N. (2006). Biopower today. *BioSocieties, 1*(2), 195–217. https://doi.org/10.1017/S1745855206040014

Ristovski-Slijepcevic, S., Chapman, G. E., & Beagan, B. L. (2010). Being a "good mother": Dietary governmentality in the family food practices of three ethno-cultural groups in Canada. *Health: An Interdisciplinary Journal for the Social Study of Health, Illness and Medicine, 14*(5), 467–483. https://doi.org/10.1177/1363459309357267

Rizzo, G., Borrello, M., Dara Guccione, G., Schifani, G., & Cembalo, L. (2020). Organic food consumption: The relevance of the health attribute. *Sustainability, 12*(2), 595. https://doi.org/10.3390/su12020595

Rose, N. (2001). The politics of life itself. *Theory, Culture & Society, 18*(6), 1–30. https://doi.org/10.1177/02632760122052020

Sage, C. (2003). Social embeddedness and relations of regard. *Journal of Rural Studies, 19*(1), 47–60. https://doi.org/10.1016/S0743-0167(02)00044-X

Salleh, A. (2016). The Anthropocene: Thinking in "deep geological time" or deep libidinal time? *International Critical Thought, 6*(3), 422–433. https://doi.org/10.1080/21598282.2016.1197784

Samsioe, E., & Fuentes, C. (2022). Digitalizing shopping routines: Re-organizing household practices to enable sustainable food provisioning. *Sustainable Production and Consumption, 29*, 807–819. https://doi.org/10.1016/j.spc.2021.07.019

Sassatelli, R. (2000). The commercialization of discipline: Keep-fit culture and its values. *Journal of Modern Italian Studies, 5*(3), 396–411. https://doi.org/10.1080/1354571X.2000.9728261

Sassatelli, R. (2004). The political morality of food: Discourses, contestation and alternative consumption. In M. Harvey, A. McMeekin, & A. Warde (Eds.), *Qualities of food.* Manchester University Press.

Sassatelli, R. (2015). Consumer culture, sustainability and a new vision of consumer sovereignty. *Sociologia Ruralis, 55*(4), 483–496. https://doi.org/10.1111/soru.12081

Schaefer, S. E., Biltekoff, C., Thomas, C., & Rashedi, R. N. (2016). Healthy, vague: Exploring health as a priority in food choice. *Food, Culture & Society, 19*(2), 227–250. https://doi.org/10.1080/15528014.2016.1178527

Schneider, T., & Eli, K. (2022). The digital labor of ethical food consumption: A new research agenda for studying everyday food digitalization. *Agriculture and Human Values*. https://doi.org/10.1007/s10460-022-10390-7

Shiva, V. (1988). *Staying alive: Women, ecology, and survival in India*. Kali for Women.

Shugart, H. A. (2014). Food fixations: Reconfiguring class in contemporary US food discourse. *Food, Culture & Society, 17*(2), 261–281. https://doi.org/10.2752/175174414X13871910531665

Sikka, T. (2019). The contradictions of a superfood consumerism in a postfeminist, neoliberal world. *Food, Culture & Society, 22*(3), 354–375. https://doi.org/10.1080/15528014.2019.1580534

Smith, J., Kostelecký, T., & Jehlička, P. (2015). Quietly does it: Questioning assumptions about class, sustainability and consumption. *Geoforum, 67*, 223–232. https://doi.org/10.1016/j.geoforum.2015.03.017

Southerton, D., & Fuentes, C. (2022). Special section: Digital platforms and sustainable food consumption transitions. *Sustainable Production and Consumption, 29*, 805–806. https://doi.org/10.1016/j.spc.2021.08.023

Spaargaren, G., Oosterveer, P., & Loeber, A. (Eds.). (2012). *Food practices in transition: Changing food consumption, retail and production in the age of reflexive modernity*. Routledge.

Vonthron, S., Perrin, C., & Soulard, C.-T. (2020). Foodscape: A scoping review and a research agenda for food security-related studies. *PLoS One, 15*(5). https://doi.org/10.1371/journal.pone.0233218

Warde, A. (2015). *The practice of eating*. Polity Press.

# A Plate Full of Food and Care: Investigating Care Practices and Gender Relations at the Crossroads of Sustainability, Innovative Practices, and Digital Platforms

*Suse Brettin, Meike Brückner, and Natalia Magnani*

## 1 Introduction

'Feeding work' (DeVault, 1991)—work of provisioning, preparing, and disposing of food—is still predominantly done by women. Even though food preparation and cooking are becoming increasingly trendy and more and more men want to participate in these jobs (Cairns et al., 2010), the gendered division of labour in households has not fundamentally changed

S. Brettin (✉) • M. Brückner
Thaer-Institute of Agricultural and Horticultural Sciences, Humboldt-University Berlin, Berlin, Germany
e-mail: suse.brettin@gender.hu-berlin.de; meike.brueckner@hu-berlin.de

N. Magnani
Sociology and Social Research, University of Trento, Trento, Italy
e-mail: natalia.magnani@unitn.it



A. Dulsrud, F. Forno (eds.), *Digital Food Provisioning in Times of Multiple Crises*, Consumption and Public Life,
https://doi.org/10.1007/978-3-031-46323-5_9

in recent decades (Budlender, 2010; Power, 2020). At the same time, the demands on food-related care work have changed significantly. In view of the discussion on sustainability, a shift in the necessary food-related care activities can be observed (Busa & Garder, 2015).

However, especially the last aspect of the new demands of food-related care work mostly challenges women to reinvent their activities in the kitchen. Feminist scholars have already criticised these new challenges in numerous ways. Weller (2017), for example, warns of a feminisation of the responsibility for sustainability. Little et al. (2009) show in their study on alternative food networks that it is especially women who (have to) spend the extra time, knowledge, and skills to provide themselves and their households with appropriate food. However, the impact of new technologies and the digitalisation of food provisioning on women's workload and the possibilities to shift and rearrange gender orders within the household are rather new topics in the discussion. There is some evidence that incorporating new technologies into households is not a smooth and straightforward process (Hargreaves et al., 2018). In particular, the integration into domestic food practices seems to require additional attention and work at first (Heidenstrøm & Hebrok, 2022; Samsioe & Fuentes, 2022)—a task that according to some empirical evidence seems to be gendered as well (Čajić et al., 2022).

Another issue lies in the fundamental understanding of these activities as private affairs. Regarding household food provision, the associated work and its implications are often seen as private negotiations far removed from the sociopolitical sphere—a notion repeatedly under scrutiny, especially from a feminist perspective (Waring & Steinem, 1988; Bock & Duden, 1997; Bauhardt, 2018). Sevenhuijsen (2003), for example, points out that care work—such as domestic food work—is very much linked and connected to and embedded within political processes. Although experiences of caring have often been left out of politics, it is clear that care activities are highly influenced—both directly and indirectly—by public policies such as family and health policy or labour regulations (Sevenhuijsen, 2003). Considering this influence directs the perspective to the structural side of (food-related) care work and highlights that care in general cannot merely be located in the private or with the individual.

Therein lies the starting point for criticising a focus on sustainable or ethical consumption from a feminist perspective: because these approaches mainly remain entrenched in the private and individual, they run the risk of shifting the responsibility for sustainability further and further to the detriment of women. In the following pages, we take up this discussion and expand it by considering the influence of new technologies, especially the influence of digitalised food provision on food-related care work. We will analyse to what extent digital services of food provision enable or change food-related care work. Do they represent potentials towards more sustainability as well as towards more just gender relations in household food practices?

## 2 Food, Care, Technology, and Gender: Theoretical Considerations

### *Caring Politics and Feminist Ecological Citizenship*

As mentioned above, when related to food, the necessary labour within households and its ethical negotiation is often seen as a private decision far from political action. However, this understanding is widely criticised—especially with regard to discussions on sustainable or ethical consumption. Morgan (2010), for example, argues for a new politics of care and cautions that a differentiation between ethical consumerism and ethical citizenship is necessary for any changes to become politically meaningful. He argues that care needs to be 'acknowledged as a public discourse for citizens and not merely as a private discourse for consumers' (Morgan, 2010, p. 1863).

The concept of *feminist ecological citizenship* from MacGregor (2004, 2010) goes in some ways in a similar direction: socioecological decisions in the seemingly private context of the household are to be linked to the sociopolitical level. She thereby follows the feminist economic critique that has been taking place for decades, pointing to the need for a comprehensive recognition of care, because activities of maintaining, continuing, and repairing human and nonhuman realities of life (Tronto & Fisher, 1990, p. 40) must be regarded as fundamental for a functioning (market) economy and for society as a whole (Waring & Steinem, 1988; Mellor, 2006; Winker, 2011). By expressing profound scepticism towards care ethics as a double-edged sword, MacGregor moreover cautions us about 'an uncritical emphasis on women's care-related morality [that] can also affirm harmful assumptions about gender and reify exclusionary notions about the nature of care and indeed, of carers' (2004, p. 61). Acknowledging the vital and critical debate on ecological citizenship (Gabrielson, 2008;

Gabrielson & Parady, 2010; MacGregor, 2016), we consider MacGregor's concept in this article as a starting point to weave discussions around care more firmly into the sociopolitical discourse. Building on Bowden's (1997) argument for shedding light also on the frustrating, demeaning, and isolating sides of care, MacGregor urges us not to romanticise these practices. She argues that care—towards human and nonhuman realities, serving society as a whole—should be given the sociopolitical status it deserves and needs in order to be carried out under good conditions:

> What makes feminist ecological citizenship distinct from other approaches is that it refuses the privatisation and feminisation of care and calls for public debate and action on how foundational acts of labour (e.g., care) can be recognised to allow for women's equal participation as citizens. Care is thereby politicised as a necessary part of citizenship, rather than as a 'natural resource'; that sustains action in the public sphere. (MacGregor, 2010, p. 29)

Other authors also emphasise the workload and consider the pitfalls of care and care ethics but still argue that one can learn from the knowledge, experience, and skill (Brückner, 2020). This argument directs the perspective towards the fulfilling, agentic, and inspiring sides of care and citizenship—for example, often enacted in food collectives (see, e.g., Wember, 2019). Although there are both divergent and colliding positions regarding care and citizenship, as this section shows, both positions urge us to find ways to look at the enabling moments of care while not neglecting the downside and its entanglement in politics.

Furthermore, it has become clear that care is not bound to certain places—the household or the private sphere—but rather that it exists in and through relationships in a '*multispecies world*' (Straube, 2019; emphasis in the original) of interdependencies. Thus, care can be oriented towards social as well as ecological contexts (Tronto & Fisher, 1990). In this perspective, care clearly relates to issues of sustainability, because it is about preservation and restoration of human and nonhuman life.

This vision of a fundamentally more caring world drives the authors to look at food practices from the perspective of care. In the context of this chapter, we analyse the use of technological innovations for food provisioning with the help of a care perspective in order to discuss their potentials and challenges for socioecological care.

Looking at the food system in general, questions and issues related to the spread of technology and digitalisation are already being discussed (see, e.g., Prause et al., 2021). What is missing, however, is an analysis of the influence of technological innovations on food-related care work. Even though feminist theory is no stranger to questions of technology, this specific issue so far seems to be overlooked. Thus, in the following section, we briefly scrutinise feminist science and technology studies in order to concretise our perspective.

### *Sociotechnical Innovations of Food Provisioning and Gender*

Feminist scholars have started to analyse the gendered character of technology since the 1980s. Radical feminists argued that 'Western technology is deeply implicated in the masculine project of domination and control of women and nature' (Wajcman, 2010, p. 146) and stressed the way women are not always well served by technologies. In this perspective, artefacts were not treated as neutral. On the contrary, 'technology was seen as socially shaped, but shaped by men to the exclusion of women' (ibid., p. 147). The result of this perspective was a pessimistic view of the possibility of reshaping technologies for gender equality.

This negative account generated a reaction by feminist approaches in the 1990s. Cyberfeminists of the 1990s, such as Donna Haraway, adopted an optimistic approach considering digital technologies as blurring the boundaries between humans and machines, male and female, and thus offering new avenues for emancipation. More recently, scholars within the field of Science and Technology Studies (STS) have developed a different, more nuanced perspective, which will inform our contribution (see also Chen & Dulsrud, this volume). They conceive of technology as both a source and a consequence of gender relations (Wajcman, 2004). Accordingly, gender and technology are considered mutually shaped. On the one hand, domestic technologies are already imbued with gender scripts defining their appropriate operators and their appropriate uses. On the other hand, 'the social meanings of a technology are contingently stabilised and contestable' and 'people continuously interpret, appropriate and use artefacts in their daily lives' (Wajcman, 2010, p. 150). In this perspective, gender is not conceived of as fixed and existing independently of technology but as performativity, as the product of a changing relational process (ibid.).

Specifically in the area of food-related care work, new technologies, such as smartphones or technical equipment for storing, preparing, or cooking, transform cooking practices and knowledge (Graf, 2022). These technologies have been considered to reduce the amount of work and knowledge needed for care-related food work; however, studies indicate that neither a simplification nor a fundamental shift in gendered practices has occurred (ibid.; Meah & Jackson, 2013). Nevertheless, digital technologies have the potential to 'widen younger Marrakechi cooks' relations and their sources of knowledge' (Graf, 2022, p. 17), as Graf shows in her ethnographic research in Marrakech.

Consistent with this view, in our contribution we will explore the complex relationship between gender, caring, and technology by scrutinising how gender is performed and shaped together with technology in the making. More specifically, we will ask to what extent the technical restructuring of food provisioning reshapes food-related care practices: To what extent is care work in its social and ecological dimension enabled or even changed and expanded through digital channels of food provision? Do they even represent a potential towards more just gender relations in household food practices?

## 3 Empirical Material and Contexts

Our study focuses on two case studies of urban contexts, namely Berlin and Milan. They have been chosen as case studies because they form an interesting environment in relation to food infrastructure and policies. They present a mix of differences and similarities that make a comparison between them particularly challenging. On the one hand, the regional and national contexts in which they are based present many differences. In particular, differences concern the online food provisioning systems of Italy and Germany (PLATEFORMS WP1 Public Report). Germany, from 2015 to 2018, has seen a very rapid and significant growth in online food platforms, less so in Italy, where the growth has been much less pronounced. Moreover, whereas in Italy the most diffused form of online food provisioning services (OFPS) is online only services (57.4%), i.e., those retailers that exist only as digital stores, in Germany, it is independent stores (58%), i.e., websites of small food boutiques and grocers. Almost all (91%) Italian OFPS have a national/international market scope, while in Germany, most OFPS are shipping locally. Looking into the value commitments of the OFPS in the two countries, what emerges is that a commitment towards the environment is much higher in Germany (58%) than in Italy (21.8%). Finally, the percentage of online food provisioning services selling just organic products is much larger in Germany (60.24%) than in Italy (8.53%).

On the other hand, Milan and Berlin also have many similarities. In particular, both cities are in various ways committed towards establishing a sustainable, inclusive, and resilient food system. Both cities signed the Milan Urban Food Policy Pact. Furthermore, a wide range of alternative food projects are based in Berlin: The largest number of members on Foodsharing.de (Morrow, 2019) and the largest Marktschwärmer community in Germany are active in Berlin. Moreover, there exist diverse consumer-led or consumer-involved initiatives, such as community-supported agriculture (CSA) or Food Coops. At the same time, Milan has a Food Policy Council, the largest number of solidarity purchasing groups (Gruppi di Acquisto Solidale, GAS), and the biggest Food Assembly ('Alveare che dice si') community. It can thus be observed in both cities that the demand for locally produced and manufactured food has increased, and that a variety of localised supply chains have emerged like regional fruit/vegetable boxes, solidarity purchasing groups, and Food Coops, as well as Food Assemblies. Some of these infrastructures have already existed for some time but have undergone a change due to digitalisation particularly within the last years in how they communicate with members and customers, in organising the ordering and delivery logistics, or in offering networking options amongst members.

In the case of Milan, a total of 45 households have been interviewed for the study, of which 24 (18 women and 6 men) are considered in this chapter. For the purpose of this chapter, the following three digital food platforms have been included: *Alveare che dice sì* (Italian Food Assembly), *Buonmenrcato*, and *Cortilia*. *Alveare che dice sì* has identical operations and functions to those of the *Marktschwärmer* (German Food Assembly) in Germany. It is organised around collection points where consumers meet local producers who deliver preordered food items. *Buonmercato* is a not-for-profit association for local, direct, and fair food provisioning based on the periphery of Milan. It was founded 10 years ago as a product of the GAS movement. However, unlike GAS, it allows individuals to purchase food online, via a website, and in small quantities. Orders are placed by Monday, and people collect them every week on Fridays and Saturdays at the association's headquarters. *Cortilia* is a business-driven platform offering a delivery service that buys from local producers and provides both box schemes and items on demand.

In the case of Berlin, a total of 40 interviews have been conducted for the study, 24 (7 men and 19 women due to two interviews to heterosexual couples) of which are analysed for this chapter. For the purpose of this chapter, the following platforms have been included: *Marktschwärmer*, community-supported agriculture (CSA), and regional fruit/vegetable boxes. CSA connects producers and consumers locally. Typically, CSA members share certain risks with the farms they receive the harvest from, for example, in case of financial difficulties or when work support is needed on the farm. Consumers buy their share of the harvest up-front and become members for a year, which gives producers security and allows them to focus on farming activities. In most cases, digital tools and technologies support CSA, for example, in sharing information or organising orders. Fruit and vegetable boxes are offered by several enterprises in and around Berlin. The ordering is conducted digitally as well as the communication, yet the box is delivered straight to the customer's door. The characteristics of the boxes vary: Some deliver products produced seasonally and regionally, others offer a wide range of products from different suppliers and production regions.

## 4 Analysis of Food-Related Practices

To find out to what extent the integration of OFPS has changed or restructured food-related care work in households, we will first look at the reasons and motivations—why did practices of food provision change in the first place? What were the reasons for introducing OFPS into household food practices? We first look within food-related care relations and food-related caring contexts in the household and then turn to food-related care relationships beyond the household. Subsequently, we scrutinise the specific changes of food practices both in their layout and in their organisation. First, we analyse whether the here-considered OFPS change the gender workload and thus to what extent they can favour a more gender-balanced division of labour. Second, we focus on the effects of these OFPS on relationships between consumers and producers.

### *Changing Household Food Practices: Care Within the Household*

In the analysis of our data material, it very quickly became clear that changes in food provision are very often related to changed demands and relationships regarding care. In both cities, we found numerous reports of changed circumstances regarding care within households, changes that caused the integration of new channels of food provisioning. Not surprisingly, the most explicit shifts in practices of care we have encountered in interviews are in households with children. In these cases, food-related care work very often not only had to meet the needs and wishes of the children but also played a vital role in the very often self-imposed claims of the carer. The quote from MZ, a mother who lives with her son in a shared flat in Berlin, illustrates these two aspects very well:

> But at the same time, I have a stronger feeling that I really want to cook fresh and myself all the time. Especially because I have a child now. For me, I can also quickly eat a sandwich in the evening. I don't mind that, we do it sometimes when it's not possible, but on the whole, it's become even more important to me to cook fresh and myself, because it's not just for me. (MZ, female, CSA subscriber, Berlin)

Caring for herself alone, MZ would place less emphasis on food provision. However, because she has the additional responsibility of caring for her child, her food-related demands have increased. Later in the interview, she explicitly mentioned that she appreciates and values the food even more if it is sourced regionally and grown ecologically. For her, these parameters are an important basis for a positive evaluation of her overall food practices.

Particularly in the interviews in Milan, the birth of children and the circumstance of raising them in a context of perceived food scares were identified as key points for changing ways in food provisioning. NA, a mother, reported how she started buying through Buonmercato:

> My daughter was 5–6 years old, and at that time, in Milan, there was a big controversy about hormones and because my relatives had a cow, I started telling them: 'why don't share the meat?' [...] Moreover, I started going to Buonmercato to get vegetables to make vegetable broth for my daughter because I was against baby food. (NA, female, Buonmercato consumer, Milan)

Thus, although NA's reported motivation to shop at Buonmercato relates to changes due to an ongoing social controversy, she connects this motivation directly to her responsibility to care for her child.

However, changes in household food practices were not reported only because of children. In one case in Milan, increased health requirements due to ageing or physical ailments were given as reasons. Food then becomes a crucial means to heal the body as an alternative to medicines, as described by MI, who also shops at Buonmercato:

> Becoming older there are also health problem [...] My husband suffers from diabetes and high blood pressure. He has to eat certain foods. I suffer from reflux [...] rather than taking pills, we prefer to stay on a diet and avoid junk food [...] so our way of eating has changed a lot. (MI, female, Buonmercato consumer, Milan)

Beyond the material provision of good and healthy food, concerns about intangible aspects of food provision also played a role in both case studies. In both places, changes in household food practices, and especially changes in channels of food provisioning, were reported to be an expression of care towards more ecological knowledge or towards a certain style of consumption and eating. A very vivid example of such expressions of care is the following quote from GK, in which she recollects why she started subscribing to a regional fruit and vegetable box:

> I think the reason for the box was more that you somehow don't really have an overview of what's available in season any more. Especially with the children, but also for us, for all of us, because you actually get everything the whole year round. And at some point, we also had a calendar where you could see, or in the kindergarten, there is also a list of what fruit is available in November or December, for example. (GK, female, fruit/vegetable box subscriber, Berlin)

GK chose a box because she wants to give her children—and in a way herself—some sort of ecological education—an understanding of what agricultural products with a spatial-temporal embeddedness are even possible to be consumed. In this sense, the desire to educate her children about household food provision also points *beyond* the household, because the goal is about understanding one's own socioecological embeddedness.

However, in both Berlin and Milan, the changes in food provisioning due to responsibilities and relationships of care within the household seem to mainly involve food quality or what is bought. The following quote from an interview in Milan is thus rather exceptional in that it addresses a different aspect of the decision to change food practices and start using a digital provisioning channel:

> I tried to go shopping with the solidarity purchasing group, but it was too complicated because I did not have enough time [...] Especially now with the grandson and the grandparents to take care of, we are more busy than before therefore we mainly oriented ourselves to online shopping platforms. (MI, female, Buonmercato consumer, Milan)

MI was in fact the only interviewee who reported an explicit decision to use a digital channel of food provision. Faced with a double burden of care for her grandson as well as for her grandparents, the technological infrastructure of online shopping is first and foremost an opportunity to save time for her. However, this time-saving aspect makes her case somewhat of an exception in the data material we collected. What she has in common with many of the other interviewees—and what has become clear in the analysis so far—is that the changes in food provision are often closely related to changes in care relationships and thus caring demands in the household. Yet, the quality and type of the food offered through certain services played a major role, and shifts motivated by these issues were frequent topics in interviews in Milan (see also Dal Gobbo, this volume) and Berlin. Aspects such as saving time or convenience, on the other hand, played less of a role and were explicitly found in only one interview.

However, it has also become clear that the parameters for assessing food supply often point beyond the household—ecological aspects such as regionality and organics, and thus also questions of production methods, played a role in many interviews and were put in direct relation to the responsibilities of care in the households. For this reason, we will focus on this topic in the following section.

### *Changing Household Food Practices: Socioecological Care Beyond the Household*

As mentioned in the previous section, ecological aspects very often were central to the decision to introduce digital channels of food provisioning into household food practices, in both Berlin and Milan. The following series of quotes is intended to be emblematic of the many ways in which these aspects were addressed in the interviews in the two research contexts:

> For me at this moment, the ethics—the where the food comes from—who is behind it—is more important than quality [...] if we want to change

something in this world even with the climate crisis, you have to think about your footprint on the planet. (NA, female Buonmercato consumer, Milan)

It is really a more ethical question, because we know better where the products come from. We know that no one is exploited to collect them. Then obviously they are better. It's all a chain. (VS, female, Buonmercato consumer, Milan)

Well, the purpose was to support the regional farmers and companies and to promote and support this idea of sustainability. And just not to throw away food and to counteract this overproduction a little bit. (SR, female, fruit/vegetable box subscriber, Berlin)

The quotes clearly show that the way the products they considered purchasing were produced played a very important role for many interviewees—they linked their decisions around household food provisioning to questions of climate change as well as to working conditions for farmers. The considerations and stories about these issues seemed to be more important to our respondents than the way in which the products were ultimately purchased—i.e., whether via digital services or not. Thus, being able to act upon a sense of socioecological care, of caring *about* the environment or farmers and thus *about* socioecological issues and climate change were quite often at the heart of the decision for introducing OFPS.

Interestingly, a difference can be seen between the interviews in the two locations: the wording for describing socioecological care. In Milan, issues of origin as well as cultivation and production methods were mostly negotiated within the frame of food ethics, whereas in Berlin, the word 'ethics' was almost never used and instead a strong reference to sustainability was made. Even if the contents of the above quotations as well as the general direction of the reports in the various interviews ultimately amount to the same thing, we find it interesting that the wording under which socioecological issues are negotiated is different—ethics and sustainability. We believe it is worthwhile to investigate this difference in more detail through further research because the interviews in the two cities also revealed certain undertones and tendencies. In Milan, interviewees were often concerned with 'doing the right thing'—acting ethically. And they seemed quite willing to make sacrifices to do so. To some extent, Berlin interviewees were also prepared to make sacrifices, but mostly they embedded their food practices into the larger food system as a whole—speaking of the limits of one's own actions and seeing oneself as part of a larger, necessary change towards more sustainability. It would therefore be interesting to further explore to what extent discussions on nature–society relations and the socioecological crisis in the two research contexts—Italy and Germany—do actually take place within different discursive framings or premises and, above all, what influence this possibly different framing has on (changing) household food practices.

This possible difference in framing also shines through when talking about another aspect of socioecological care—the interviewees' care *about* good working and production conditions for farmers. Again, this point is embedded somewhat differently in Berlin from how it is embedded in Milan, as can already be seen in some of the quotes above as well as in the following ones:

> Even the ethical question is important [...] I think it is important to support small producers. [...] I prefer to pay for fruit and vegetables what they are worth rather than going to buy things that I know are the result of exploitation. (SM, female, Alveare consumer, Milan)

> But above all, I believe that it is sustainable in the sense that the farmers have planning security. So, I think that's also something that's super important to me. I think our current economic system is super mean to people who produce food. And I think that this [CSA] is really something that simply distributes the risk over many more shoulders and somehow supports them and can also support small businesses. And that, I would say, is something sustainable. (VP, female, CSA subscriber, Berlin)

If SM from Milan talks above about an ethical decision that motivates her to support farmers, for BL from Berlin it is a question of sustainability, and she even makes explicit references to systemic justice and puts her own way of food provisioning in relation to this principle. In both cases, however, the respondents clearly feel a strong responsibility towards farmers and do care *about* their living and working conditions. However, because few city dwellers have direct links to farms to express their concern directly, it is ultimately the choice of food channels that allows them to respond accordingly. At this point, the limitation of individual responsibility, and especially of individual solutions, becomes obvious. To think further, the analysis of different relationships of care done by Noddings (2010) proves to be helpful because she distinguishes between *caring for* and *caring about*. The first relationship is a direct one, which is characterised by actual encounters and explicit actions, i.e., concrete care work for a child. The second relationship of care relates to the needs of people with whom we do not have direct encounters (Noddings, 2010, p. 22)—in this case, the working and production conditions of farmers. Especially in the perspective of *caring about*, it becomes clear that individual and private solutions cannot do justice to this relationship of care—towards people who are not part of one's everyday life. Noddings formulates this limitation as follows:

> Because it is impossible for any human being to care for everyone, we must find a way to care about the problems of people we cannot reach individually. I've suggested that this should be a collective enterprise, in part to achieve some stability in giving and, in part, to relieve individuals from feeling overwhelmed by the many demands on their desire to care. (Noddings, 2010, p. 25f)

In this perspective, infrastructures and channels of food supply can thus be understood as a form of care—socioecological care about the environment and people working within the food system. In an increasingly complex food system, it cannot be the challenge of individual households alone to develop those structures, and as Noddings suggested, it should rather be a collective effort to promote those infrastructures that enable people to act upon their *care-about* desire.

Another aspect related to socioecological care that we furthermore observed in various interviews, especially in Berlin and that ties in with Nodding's analysis as well, is the feeling of being overwhelmed. Since food provision is foremost about satisfying the different needs within a household, many respondents reported about the demand to find a balance between individual taste, pleasure, well-being, and socioecological care. Using specific services of food provisioning thus helped them to maintain this balance as the following women tell us:

> And in principle we do that because they really say: Here, this is the cow, we got it from this farmer, it's there. He tells us that everything is species-appropriate in terms of husbandry. And the cow is sold in its entirety and then shipped. It's not like in the supermarket, where half of it is thrown away because no one bought it, but because it's more sustainable. We also eat less meat, although the boys would want to eat meat all the time. But still, we all want it now and then, so we do buy it, but then it's better. (CK, female, fruit/vegetable box subscriber, Berlin)

> I look for a balance between attention to sustainability and the fact that what we take for ourselves has an impact on others and we are lucky that they go together quite a lot. If you take care of your diet, you are doing good for yourself and for the world. (IP, female, Alveare consumer, Milan)

As is very well illustrated in the quotes, socioecological care requires constant negotiation between different needs and demands—these relate both to members of the household and to needs and demands beyond that. Especially in regard to the latter, many respondents reported struggles to respond to the various demands placed on food-related care work and of feelings of being overwhelmed. GK, for example, talks about her time before subscribing to a regional fruit and vegetable box:

> I'd rather be more consistent, but that's what we were talking about at the beginning, that I have to make a choice, that is price, organic, availability, how I can do it in my time frame. And then regional, that's just really difficult. And in the end, it's like that. I tried it for a while, but then I really despaired sometimes, because sometimes I didn't know what to buy anymore. (GK, female, fruit/vegetable Box subscriber, Berlin)

For GK, food provision via a digitally ordered fruit and vegetable box is therefore an enormous relief. She no longer has to weigh up the various demands—organic or regional production, price, seasonality—with every single purchase decision, and the feeling of being overwhelmed largely disappeared. Consequently, new modes of food provision discussed in this article offer an enormous potential because they facilitate the inclusion of socioecological care in everyday life. We call the relief that comes with the OFPS used in our data *socioecological convenience*. This term refers to convenience, which promises relief not so much in time or labour savings, but above all in making complex sustainability decisions. Users of these OFPS—Buonmercato and Alveare in Milan; Marktschwärmer, CSA, and regional fruit and vegetable boxes in Berlin—do not have to reevaluate every purchase decision, but opt for a certain service, which in turn promises to meet their demands regarding socioecological care and thus sustainability.

We therefore argue that within the scope of cases we studied, changes in food provision are primarily based on relationships of responsibility and care, rather than on an explicit desire to save time or on other aspects of what is typically understood as convenience in relation to food. Changes towards digitalisation were therefore mostly a side effect in our sample and not the motivation or reason for them. Accordingly, in both Berlin and Milan, the desire to explicitly introduce OFPS was less prominent. Rather, changes in food provision were motivated by the desire for *careful* socioecological provision structures in relation to both the household and beyond.

But if the desire for shifting workloads or changing material and immaterial resources is not the motivation for introducing the channels, how are these aspects related? Questions around workloads—in particular from a gendered perspective—will therefore receive greater attention in the following section. We ask what exactly the OFPS have changed in households?

### *Changing Gendered Workloads*

As already mentioned, data indicate that women internationally still manage the majority of food-related care work within households, as was also reflected in our interviews. On the one hand, it was mainly women who made themselves available as interview partners on the topic of household food practices, even though we addressed the household as a whole. Moreover, it was mainly women who introduced and integrated new channels of food provision in the households. CG, a regular shopper at Marktschwärmer in Berlin, for example, tells of her efforts, which are the basis for using this channel:

> So, we are trying to eat less meat, less chicken, less fish, but when we eat them, we want them to be good quality, coming from good sources and where the animal has grown and was fed properly. Other than that, I am more into buying locally. [Name of her partner] is way less than that. He is less keen on that than me, so I try to shift the household in that direction. (CG, female, Marktschwärmer consumer, Berlin)

It is CG whose socioecological *care-about* attitude is the starting point for using Marktschwärmer—she is the one who tries 'to shift the household in that direction' and introduces a new—digitised—channel of food provisioning. Nonetheless, as already mentioned, adopting new channels of food provision is not a seamless and effortless process, especially when these channels require integrating new technologies (De Bernardi et al., 2019). Again, in our interviews, it was often the women who took on this effort. For example, in the following quote, MZ reports on the tension between the benefits and costs of OFPS:

I: Do you also use the [CSA] as a source of exchange for recipes or how to prepare something?

R: Not yet, but I often think about it. I'm just not yet so into it that I can keep up with the collection and yes, so my, let's say, common exchange channels are the easiest for me at first. Starting new things again now just takes a bit of time, because I simply don't have the time in the evenings or so to sit down on the internet, deal with this SoLaWi website again and then put in or get out a bit more. (MZ, female, CSA subscriber, Berlin)

Even if the CSA's digital offer for exchange and knowledge transfer could definitely be an enrichment, for MZ it is first and foremost an additional effort. In interviews in both Milan and Berlin, it became clear that digital channels of food provisioning have this ambivalent impact on workloads in general and on the gendered division of food-related care work in particular. In most cases, a traditional division of workloads is reestablished: Women shop online and men pick up the order when they get back from work. But there are some exceptions in which the ordering is done together.

> We do the order together seated on the sofa and 99% of the time he passes by to pick it up because he likes doing that, it becomes a moment for socialisation. (VC, female, Buonmercato consumer, Milan)

Stories like this one, of ordering together—whether from the sofa or even in the kitchen with a direct view into the fridge and the pantry shelf—appeared sporadically. In one interview in Berlin, it was even reported that the entire family regularly came together and discussed the food for the coming days while the order was being placed online. Especially against the background of the challenge of balancing different needs and desires within a household, as described above, the possible relief provided by this change in the process of food planning and provision becomes obvious.

Within the framework of the joint analysis of the interviews, however, it became clear that reports of joint digital ordering and thus also of joint planning processes in relation to food were mainly found in the Berlin interviews. We interpret this as meaning that digital services certainly offer the potential for changing the gendered division of labour, but that this potential is still embedded in the respective, overall social context of the gendered division of labour. A closer look at international time-budget studies can be very informative in this regard because these studies show differences between the gendered time-use patterns in Italy and those in Germany. In particular, in Italy the average time spent by women in household and family care is higher than in Germany—in the last Harmonised European Time Use Surveys (HETUS) from 2010, time frames of 5 hours (Italy) versus 3.50 hours (Germany) are listed. Men in Italy spend less time in household and family care (1.39 hours in Italy versus 2.17 in Germany). Also, time spent by women in food management is tendentially higher for Italy than for Germany (1.26 hours vs. 44 min). Generally speaking, women in Italy spend more time on household and family care work than women in Germany do; men in Italy, on the other hand, spend less time on care work than men in Germany do. Linking these findings back to our analysis, we would argue that although digitalising food provision does offer potential for facilitating certain work steps, the basic gender division of labour will not change fundamentally. With reference to STS theories, however, this persistence comes as little surprise because technology in this perspective is both the source and the consequence of gender relations. Subsequently, it would certainly be illuminating to analyse in more detail the extent to which certain technologies and modes of digitalisation in relation to food practices are coded female or male, i.e., whether they appeal more or less to certain groups of users. It is to be expected that gender as a social structural category and ordering principle also plays a decisive role here.

On the contrary, there were also indications in various interviews in both Berlin and Milan that introducing digital channels of food provisioning has increased the workload of women in particular. Especially the digital ordering of vegetable boxes or being a member in a digitally supported solidarity purchasing group reorganises food-related care work quite often to the detriment of women because these modes of food provisioning strongly affect storage, preparation, and cooking. The fruit and vegetables—usually coming fresh from the field—must be cleaned, properly stored, and preserved; inventiveness and flexibility are necessary for preparation. The following two quotes from both cities are thus exemplary and illustrate very well the challenges just described:

> There have been times when I exaggerated buying fruit and vegetables. I then found myself with all kinds of vegetables to cook and then I made 5 packs of vegetables to put in the freezer [...] when it arrived it was wonder-

> ful because then I really like vegetables and also fruit but then I had to cook all evening, washing the spinach, washing the herbs, preparing them [….] (AC, female, Alveare consumer, Milan)

> In the box, I just get it in front of me and then I'm just happy that I can use it. Then I first look: What am I going to do with it now? I don't know what I can cook with it now, what I can do with it at all. And then it's just the other way round. Otherwise, you have this recipe in your head and you buy for it, and that's how it is, you are presented with things and then you look for a recipe. And that's actually always quite nice. (GK, female, fruit/vegetable box subscriber, Berlin)

At the same time, it was precisely the surprise effect of the vegetable boxes—despite a possibly higher workload—that was appreciated in Milan. This effect gave interviewees the opportunity to experiment with new foods and ingredients as well as to get to know new vegetables and recipes.

> I really like this thing of not knowing what's coming to you. The double value is that the producer throws out the stuff that would go bad if nobody buys [it]. And the other value is that the consumer buys things that he would never really buy. I have discovered stuff that travels only through this channel […] in supermarkets you cannot find things like mustard leaves or white turnips. I had never eaten them before. (DC, female, Buonmercato consumer, Milan)

> Our curiosity has also changed a bit. Because before, if we went to the coop, we always took the same things of fruit and vegetables. Instead, at Buonmercato there are also some things a bit special, at least for us, and we are still curious to take them and therefore this involves to experiment with things we had never tried. (VS, female, Buonmercato consumer, Milan)

The analysis of the data thus does not reveal a clear picture of the influence of new technologies of food provisioning on the gendered division of labour. Whereas regarding planning, there seems to be a reduction and a relief in the (mental) workload or at least a new distribution of work. In other respects, new steps of food-related care work have been added (see also Dal Gobbo, this volume). However, as the last quotes show, these new steps too are viewed ambivalently and in part are seen as an enrichment. Especially in the interviews in Milan, it almost seems as if the innovative nature of the digitised form of food provisioning is driving innovation in food preparation in the household and this innovation in turn is an essential part of the positive evaluation of these new services—despite the possibly higher workload.

But beyond workloads, what else has changed in households with the introduction of new technologies of food provisioning? We will address this question in the final section of the analysis.

### *Changing Relationships Between Producers and Consumers*

In the final section of the analysis, we will turn to the aspect that has undergone the greatest shift through the OFPS considered in this article—connections and relationships beyond the household. As discussed in the above sections, the channels of food provision considered here represent an infrastructure that enables practices of socioecological care in households. As outlined in the theoretical section, practices of care always require some form of relationship or—as in the case of care about—another person or entity. It is therefore hardly surprising that digital platforms are especially appreciated because they allow a new—direct—relationship between consumer and producer:

> [...] but it is the case that they also maintain communication with their members and also invite a lot of people out there, when it is not Corona somehow, and organise events and offer guided tours and so on and so forth. So, you can get in touch through that. That's for sure. (SB, male, fruit/vegetable box subscriber Berlin)

> And now I am grateful that I have the opportunity to buy supraregional things directly from producers, from small producers, so that I don't throw my money down the throat of some big corporation. (SR, female Marktschwärmer consumer, Berlin)

In the Berlin context in particular, the opportunity to buy directly from the producing farmers and to be able to come into direct contact with them was especially appreciated, as the two quotations make clear. But also in Milan, positive references to the possibilities of digital information acquisition in relation to products and producers could be found in the interviews, as, for example, VS states in the following quote:

> In the website for each product, it says who the producer is and then there is a description of the producer if it is organic or not, if the package is in

> kilos, etc. Then I very much appreciate that they [Buonmercato] also organise visits to the various producers and so we it happened that we had lunches with the producers. (VS, female, Buonmercato consumer, Milan)

In our opinion, the basic appreciation of the opportunity to have or make contact with producers can be directly linked to socioecological care. As discussed, the channels of food provision considered here represent the 'collective enterprise' (Noddings, 2010) through which people are enabled to care for distant others—human or nonhuman. This infrastructure, and thus relationality, of course, is shifting the most through digitalisation because it enables new, almost everyday forms of relationships, especially for urban households. This reorganisation of the relationship between consumers and producers allows an integration of a socioecological care-about attitude into households, an attitude that seems to be even more oriented towards the (assumed) needs of farmers, as the following quote from Berlin shows quite well:

> Well, I'm with the Food Assembly because it's just the coolest thing to order food. You do the regional farmers a favour because they don't have to stand in a market all day, but they rather know exactly what they have to harvest, or what kind of animal has to be slaughtered, and they have hardly any surpluses or none at all that have to be thrown away. You have a face to your products and so on. So, I think that's so cool. (AS, female, Marktschwärmer consumer, Berlin)

For AS, ordering food online directly from a farmer is 'the coolest thing' and the reasons she gives for this evaluation are all oriented towards socioecological issues: Direct ordering gives the farmer more planning security, which in turn means less food waste. In addition, the digital processing of purchases is seen as a labour-saving measure for the farmer.

The reorganisation of the relationship between producers and consumers through digital channels of food provision thereby entailed new possibilities for obtaining information and making contacts, both of which seem to be the reasons for trust in the quality of the products. One interviewee even went so far as to say that she did not pay attention to labels or certificates when ordering from certain producers, but that the mere fact that it was that particular producer—whom they collected information about before starting to order—was a guarantee of the desired quality for her:

> So yeah, I try to, you tend to get used to them, to the same farmers. You know what he is gonna have and it's a habit you develop. And it's ok. [...] This one, he has a Bio certification. But at the moment I don't care, I mean it's good that it's organic but if it was not organic, I still think I'd rather buy from him. (CG, female, Marktschwärmer consumer, Berlin)

CG orders from a number of specific farmers and therefore usually knows roughly what their product range has to offer. She does not necessarily attach importance to the certifications, but rather trusts the sometimes quite detailed descriptions online as well as—as she told us later in the interview—the possibility to see for herself on site. It thus can be argued that in our cases, digitalisation presents an opportunity to establish relationality and trust—core aspects of caring relationships in general.

What became clear though, not least in the conversation with CG, is the enormous importance of real interactions and meetings. As much as the digital information possibilities create a foundation, it is mostly the possibility for meetings in person that is often seen as a central part of the positive evaluation of certain channels of food provision:

> Buonmercato is more on a human scale ... even if the purchase method is online but it is one [a method] to take advantage of the internet technology, to experience a type of shopping on a human scale ... there are also more or less faces you know, you can meet producers, it creates a pleasant environment to hang out. (VC, female, Buonmercato consumer, Milan)

> So, we went to Potsdam and harvested carrots and potatoes last autumn. I mean, on the one hand, it was a really nice day, and it was a lot of fun. But on the other hand, it was also really nice to see where our food actually comes from. And that was already, I mean, they always send pictures and then somehow a little story or a little report about what is somehow easy to grow and what somehow causes problems and so on. But it was still really cool to meet the farmer and other people from our SoLaWi. And also, to somehow become more a part of it. (VP, female, CSA subscriber, Berlin)

Thus, digitalisation can function as a kind of door opener—overcoming the first hurdle. But as can be seen here in the quotations, many interviewees particularly appreciated the opportunities to actually get to know and meet the producer, as well as the exchange with other users and consumers. Those meetings seem to form a vital part of the digital food provision experience especially in the Milan cases of Buonmercato and the Berlin cases of CSA and Marktschwärmer. In our opinion, this shows that socio-ecological care-about attitudes can very well be promoted via digital infrastructures, but ultimately, real interactions with the other are vital and represent a fundamental component.

In conclusion, however, it should be noted that especially regarding shaping new relationships, the OFPS considered in our sample represent an opportunity. These channels create—technologically supported—connections and infrastructures within the food system that probably would not exist otherwise, at least not in this direct form (see also Dal Gobbo, this volume). Following our analysis, these technological infrastructures represent a basis for including socioecological care in households' food practices. Nevertheless, the analysis of the interviews also showed that integrating these channels, as well as the associated changes in food practices, often means additional work which is mostly gendered as well. The evaluation of this workload, however, was very different. Here, further research would certainly be profitable, especially to answer the question of how to increase the attractiveness of such channels more broadly.

## 5 Discussion and Conclusion

We have analysed the use of certain types of digital channels of food provisioning in order to investigate their potentials and challenges for practices of socioecological care within and beyond the household. We have tried to answer the following questions: To what extent is care work in its social and ecological dimension enabled or even changed and expanded through these channels? Do they represent a potential towards more just gender relations in household food practices? The results from our empirical analysis are contradictory. On the one hand, it has emerged that care clearly plays a key role in the transformation of food practices towards more sustainability within and outside the household. As our analysis shows, changes in food practices are often initiated for reasons related to care practices and caring relationships—first of all, within the household but also beyond—rather than by an instrumental motivation to save time or money. The technologically enabled innovations in food provision in our sample often seem to offer further support and enable care work towards members of the family and household, as well as towards other actors in the food chain, particularly small farmers. Especially the aspect of enabling caring relations within the food chain is linked to the fact that the

OFPS considered in our cases have a strong relational potential because they allow the reorganisation of relationships between producers and consumers by creating new (digital) opportunities and spaces for connection. The infrastructure thus created offers a basis for socioecological care-about attitudes. Following the analysis of different care relationships—*caring for* and *caring about*—Noddings develops the idea of a network that enables relationships of responsibility and care in various constellations without placing the sole burden on individuals: a 'web of care' (Noddings, 2010) understood as a collective enterprise to create viable and sustainable care relationships. The OFPS considered in this article can thus be described as such a 'web of care' (Noddings, 2010) regarding food provision. However, acknowledging McGregor's (2010) argument for a stronger politicisation of care, providing this '*web of food provision*' should be understood not only as a collective enterprise but also as a socio-political task—enabling households to care (and thus consume) in a more sustainable way, without shifting the extra effort onto the shoulders of women, who still do the majority of food-related care work. Following this line of argumentation, food practices could then be thought of as part of *ecological citizenship* and less as an individual challenge of each household towards more sustainable consumption.

Finally, sustainability does not take place in a vacuum but instead manifests in the context of concrete practices—socioecological care—which are carried out by people—in the context of food, mostly women. The subsequent overall question is thus how to foster the integration of socioecological care more strongly and equally into everyday practices as well as how to facilitate the care that is already taking place. What we were able to show through analysing our data is that our interviewees have already decided to exercise socioecological care and thus to meet certain demands within their households and beyond. Nevertheless, they very often reported a feeling of being overwhelmed by the various sustainability requirements or feeling limited in their scope of actions. We therefore consider it necessary to develop infrastructures and networks—*webs of food provision*—that minimise this feeling of being overwhelmed. As we have been able to show, the OFPS in our case studies represent a form of *convenient sustainability* and the reorganisation, to a certain extent, of relationships that facilitate the integration of socioecological care into everyday food practices.

## References


Bauhardt, C. (2018). Nature, care and gender. Feminist dilemmas. In C. Bauhardt & W. Harcourt (Eds.), *Feminist political ecology and the economics of care. In search of economic alternatives* (pp. 16–35). Routledge.

Bock, G., & Duden, B. (1997). Arbeit aus Liebe—Liebe aus Arbeit. Zur Entstehung der Hausarbeit im Kapitalismus. In Gruppe Berliner Dozentinnen Frauen und Wissenschaft (Ed.), *Beiträge zur Sommeruniversität der Frauen im Juli 1976* (pp. 118–119). Courage.

Bowden, P. (1997). *Caring: Gender-sensitive ethics.* Routledge.

Brückner, M. (2020). *Biodiversity in the kitchen. Cooking and caring for African indigenous vegetables in Kenya: A feminist approach to food sovereignty.* Oekom Verlag.

Budlender, D. (2010). *Time use studies and unpaid care work.* Routledge.

Busa, J. H., & Garder, R. (2015). Champions of the movement or fair-weather heroes? Individualization and the (a)politics of local food. *Antipode, 47*(2), 323–341. https://doi.org/10.1111/anti.12108

Cairns, K., Johnston, J., & Baumann, S. (2010). Caring about food. Doing gender in the foodie kitchen. *Gender & Society, 24*(5), 591–615.

Čajić, S., Brückner, M., & Brettin, S. (2022). A recipe for localization? Digital and analogue elements in food provisioning in Berlin A critical examination of potentials and challenges from a gender perspective. *Sustainable Production and Consumption, 29*, 820–830.

De Bernardi, P., Bertello, A., & Venuti, F. (2019). Online and on-site interactions within alternative food networks: Sustainability impact of knowledge-sharing practices. *Sustainability, 11*(5), 2–14.

DeVault, M. L. (1991). *Feeding the family. The social organization of caring as gendered work.* The University of Chicago Press.

European Union: Harmonised European time use surveys (HETUS). (2010). Eurostat. Retrieved from https://ec.europa.eu/eurostat/web/main/data/database

Gabrielson, T. (2008). Green citizenship: A review and critique. *Citizenship Studies, 12*(4), 429–446.

Gabrielson, T., & Parady, K. (2010). Corporeal citizenship: Rethinking green citizenship through the body. *Environmental Politics, 19*(3), 374–391.

Graf, K. (2022). Cooking with(out) others? Changing kitchen technologies and family values in Marrakech. *The Journal of North African Studies*, 1–26. https://doi.org/10.1080/13629387.2022.2056448

Hargreaves, T., Wilson, C., & Hauxwell-Baldwin, R. (2018). Learning to live in a smart home. *Building Research & Information, 46*(1), 127–139.

Heidenstrøm, N., & Hebrok, M. (2022). Towards realizing the sustainability potential within digital food provisioning platforms: The case of meal box

schemes and online grocery shopping in Norway. *Sustainable Production and Consumption, 29*, 831–850.
Little, J., Ilbery, B., & Watts, D. (2009). Gender, consumption and the relocalisation of food. A research agenda. *Sociologia Ruralis, 49*(3), 202–217.
MacGregor, S. (2004). From care to citizenship: Calling ecofeminism back to politics. *Ethics and the Environment, 9*(1), 56–84.
MacGregor, S. (2010). Earthcare or feminist ecological citizenship? *Femina Politica, 19*(1), 22–31.
MacGregor, S. (2016). Citizenship: Radical, feminist, and green. In T. Gabrielson, C. Hall, J. M. Meyer, & D. Schlosberg (Eds.), *The Oxford handbook of environmental political theory* (online ed., pp. 608–623). Oxford Academic.
Meah, A., & Jackson, P. (2013). Crowded kitchens: The 'democratisation' of domesticity? *Gender, Place and Culture, 20*(5), 578–596.
Mellor, M. (2006). Ecofeminist political economy. *International Journal of Green Economics, 1*(1–2), 139–150.
Morgan, K. (2010). Local and green, global and fair: The ethical foodscape and the politics of care. *Environment and Planning A: Economy and Space, 42*(8), 1852–1867.
Morrow, O. (2019). Sharing food and risk in Berlin's urban food commons. *Geoforum, 99*, 202–212.
Noddings, N. (2010). Care ethics, caregiving, and global caring. In V. Moser & I. Pinhard (Eds.), *Care—Wer sorgt für wen?* (pp. 17–26). Barbara Budrich Verlag.
Power, K. (2020). The COVID-19 pandemic has increased the care burden of women and families. *Sustainability, 16*(1), 67–73.
Prause, L., Hackfort, S., & Lindgren, M. (2021). Digitalization and the third food regime. *Agriculture and Human Values, 38*, 641–655.
Samsioe, E., & Fuentes, C. (2022). Digitalizing shopping routines: Re-organizing household practices to enable sustainable food provisioning. *Sustainable Production and Consumption, 29*, 807–819.
Sevenhuijsen, S. (2003). The place of care. The relevance of the feminist ethic of care for social policy. *Feminist Theory, 4*(2), 179–197.
Straube, W. (2019). Toxic bodies. Ticks, trans bodies, and the ethics of response-ability in art and activist writing. Environmental. *Humanities, 11*(1), 216–238.
Tronto, J. C., & Fisher, B. (1990). Toward a feminist theory of caring. In E. K. Abel & M. K. Nelson (Eds.), *Circles of care. Work and identity in women's lives* (pp. 36–54). SUNY Press.
Wajcman, J. (2004). *TechnoFeminism*. Polity Press.
Wajcman, J. (2010). Feminist theories of technology. *Cambridge Journal of Economics, 34*(1), 143–152.
Waring, M., & Steinem, G. (1988). *If women counted: A new feminist economics*. San Francisco: Harper & Row.

Weller, I. (2017). Gender dimension of sustainable consumption. In S. MacGregor (Ed.), *Routledge handbook of gender and environment* (pp. 331–344). Routledge.

Wember, C. (2019). Striving towards what we do not know yet: Living feminist political ecology in Toronto's food network. In C. Bauhardt & W. Harcourt (Eds.), *Feminist political ecology and the economics of care. In search of economic alternatives* (pp. 162–188). Routledge.

Winker, G. (2011). Soziale Reproduktion in der Krise—Care Revolution als Perspektive. *Das Argument, 53*(3), 333–344.

# Online Food Provisioning Services (OFPS) as Sociotechnical Assemblages: Three Theoretical Threads and a Proposition for Further Research

*Jiaxi Chen*

## 1 Introduction

The last two decades have witnessed the tremendous growth and expansion of digital technologies across people's everyday activities (de Reuver et al., 2018). For example, people start to use Amazon and eBay to do online shopping, Uber and Airbnb to get driving and accommodation services, respectively, and social media platforms such as Facebook and Instagram to connect with families and friends. As for the food sector, digital technologies have offered opportunities for consumers to do everyday grocery shopping in alternative ways and potentially bringing some changes in everyday food consumption. Gilliland et al. (2015) found that

J. Chen (✉)
University of Bristol Business School, University of Bristol, Bristol, UK
e-mail: jiaxi.chen@bristol.ac.uk



A. Dulsrud, F. Forno (eds.), *Digital Food Provisioning in Times of Multiple Crises*, Consumption and Public Life,
https://doi.org/10.1007/978-3-031-46323-5_10

some consumers used food-tracking applications to develop healthy dietary patterns. Schneider et al. (2019) argued that an increasing number of food campaigns were initiated and spread through social media, such as boycotts or online petitions. Although there are different kinds of digital platforms building or strengthening connections between food and consumers, it is online food provisioning services that received unprecedented attention in the COVID-19 pandemic, particularly because they fit with the lockdown and social isolation policy by delivering food straight to people's doors and reducing in-person contact (Dannenberg et al., 2020; Hassen et al., 2021).

By online food provisioning services (OFPS), this chapter refers to the concept previously introduced in Oncini's chapter (Oncini, 2024), the concept that consists of pipelines, platforms and on-demand grocery delivery that provide food for everyday consumption. Notably, even prior to the COVID-19 pandemic, OFPS had increasingly become embedded in people's everyday life, as evidenced by their diverse functions and the range of food practices they facilitated. For instance, online grocery platforms are used for everyday food consumption, which involves storing, cooking, eating and waste disposal practices (Heidenstrøm & Hebrok, 2022). Box schemes that connect consumers with local farmers are used not only because they relieve the burden of food planning, but also because they align with the values of health and sustainability (Fuentes & Samsioe, 2021; see also Fuentes et al., 2024). Therefore, given the dynamics and diversities of OFPS and their associated food consumption practices, it is important to find out theoretical approaches to conceptualise and understand how these OFPS are being used and getting embedded in people's everyday life.

To achieve this aim, I will adopt sociotechnical assemblage views in this chapter and explain why and how they can help to understand OFPS and OFPS-based food consumption. In Sect. 2, I will first explain the concept of 'sociotechnical assemblages' and the theoretical threads that it comes from, namely, Actor Network Theory (ANT), Theories of Practice (ToP) and Multi-Level Perspective (MLP). Specifically, I will discuss and compare how these three interrelated theories illuminate and inspire different aspects of sociotechnical assemblage views despite their epistemological differences. In Sect. 3, sociotechnical assemblage views will be applied to OFPS and OFPS-based food consumption, by proposing (1) OFPS as sociotechnical assemblages and (2) OFPS within sociotechnical assemblages. I aim to demonstrate why and how sociotechnical assemblage views are useful to conceptualise OFPS-based food consumption. Finally, in Sect. 4, I will discuss the potential of sociotechnical assemblage views in further research in terms of two promising themes: (1) actor interactions and agencies and (2) stability and change. These two themes also offer indispensable theoretical insights into digital and sustainable consumption research.

## 2 Sociotechnical Assemblage Views and Three Theoretical Threads

Sociotechnical assemblage views are not drawn from a single theory. They are theoretical stances shared by different theories which aim to explore the relationships between technology and society. The term 'sociotechnical assemblage' can be simply divided into 'sociotechnical' and 'assemblage'. Assemblage, according to Deleuze and Parnet (1987, p. 69), is 'a multiplicity which is made up of many heterogeneous terms...and establishes liaisons, relations between them...whose unity is that of co-functioning'. In other words, an assemblage involves multiple human and nonhuman elements and will function only when these elements are connected. The concept of 'sociotechnical' has its origin in early Science and Technology Studies (STS), whose objective is to understand technologies in their social and cultural contexts or, stated differently, the social construction of technologies (Douglas, 2012). However, these early conceptualisations tend to have a dichotomous view of technology and society, a view that consider social environment as external to technologies, providing only opportunities and constraints for societal innovation and development (Rip & Kemp, 1998, p. 338).

Inspired by these early thoughts of 'sociotechnical' and 'assemblages', three theoretical threads—ANT, ToP and MLP, arising at the turn of the twenty-first century—tried to move a step further. These theoretical attempts not only challenged the polarised understanding of early sociotechnical thinking, by exploring the intertwined and coevolving nature of technology and society, but also incorporated heterogeneous elements and their interactions into consideration when conducting these explorations. The following subsections will briefly introduce these theories and discuss how they can illuminate and inspire sociotechnical assemblage views.

### *ANT and Sociotechnical Assemblages*

ANT, at its early stage, attempts to capture how and to what extent technologies influence human behaviour and vice versa, by treating them as actors/actants in the relentless reconfiguration of society (Callon, 1999). Social and technological actors are considered interconnected, and their networks are always in the process of being created instead of finished. As multiple heterogeneous actors connect, networks are formed and a seamless field of practices is organised, which to some extent stabilises the network in return (Akrich, 1992; Latour, 2005).

ANT provides insightful views to sociotechnical assemblage thinking because 'assemblage' can be analogised to actor networks. These two terms are now often used interchangeably because of two epistemological similarities. First, both of them not only consider humans and nonhumans (such as objects, animals and nature) as constituents, but also treat human and nonhuman agencies equally (Müller, 2015). Second, both assume emergent, dynamic and connected views on networks and social phenomena, since networks and assemblages are built only when connections are made (Farías & Bender, 2012; Müller & Schurr, 2016). Thus, to understand how ANT can contribute to sociotechnical assemblage views, the features of actor networks should first be explored.

Firstly, ANT incorporates an extended range of heterogeneous entities including humans and nonhumans in its conceptual framework—actor networks. Like humans, nonhuman actors are also important actors or actants in constituting networks and assemblages. They are by no means neutral or objective, but work as agential parts that can frame human behaviour when connections are built (Latour & Johnson, 1988). ANT uses the concept of 'scripting' to describe how nonhuman entities (products, technologies, etc.) shape consumer behaviour by envisaging them to act in certain ways that are designed by the engineers and programmers (Akrich, 1992, p. 208). For example, OFPS first 'choose' their consumers as someone who owns a digital device that can connect to the Internet. In addition, different offers and promotional information also reach consumers thanks to the previously recorded purchase data and algorithmic recommendations. Then, certain steps are required for orders to be made: opening the application, choosing the product by typing its name or scrolling up and down, being directed and redirected to different categories, filling in personal information and then making payments after rejecting some of the recommendations. Also, when consumers are selecting products, it is algorithms that work out what products will be shown in what orders (Venturini et al., 2017). Based on consumers' behaviour and operation on OFPS, algorithms are able to give dynamic responses, motivating consumers' decision-making in specific ways. All these scripts are designed by technologists, with different business aims including data collection, algorithm optimisation and relentless growth (Zuboff, 2015), promoting sustainable consumption (Samsioe & Fuentes, 2022) or a healthy diet pattern (Lupton, 2021). In a sense, human actors are no longer regarded as superior to other elements. Instead, all human and nonhuman actors have an equal ontological footing, meaning that all actors equally exist and should not be ignored or erased in empirical analysis.

Secondly, ANT takes on an emergent and relational view in understanding actors and the networks as well as the phenomena they construct, a view that can be interpreted from two perspectives. On one hand, it argues that no actors have their inherent qualities and agencies, which are acquired only through connections with other actors and fulfilled in practice (Sawyer & Jarrahi, 2014). In this sense, OFPS without connections with food, producers, consumers and deliverymen cannot be considered as OFPS. For instance, normally social media are not identified as food provisioning services because their initial goal is to connect people with their friends and families. But they are now serving similar functions as OFPS by facilitating the transactions among food providers and consumers. On the contrary, although online supermarkets are recognised as OFPS without a doubt, some consumers just use them to access daily necessities such as toilet paper and shower gel (Van Droogenbroeck & Van Hove, 2017). These alternative appropriations of OFPS have encouraged a dynamic and heterogeneous view of 'OFPS'. Thus, to understand the digitalisation of food consumption, researchers need to investigate the actual use of digital devices.

On the other hand, the relational view of ANT suggests that networks are built and expanded through connections. In particular, they use the word 'translation' to describe the process of including actors into a network, which is conducted by reconfiguring and tailoring the meaning of connections to new actors, aiming to align them to fulfil a shared end (Latour, 2005, pp. 106, 128). This relational view highlights the often overlooked, but essential material entities that function backstage, such as the urban road systems, food packages and their indispensability to food delivery; fridges, room temperature and their importance to food storage; data, algorithms, interface designs and their indications to shopping behaviour (Evans, 2014; Evans et al., 2022; Fuentes & Sörum, 2019). Specifically, according to Latour et al. (2012), in the digital era and with platform media, some of the links and connections can be experienced, actualised and investigated more easily. For example, with the hyperlinks provided in OFPS, consumers can track back to the source of food providers, couriers and also the food provisioning company enabling food delivery services. Researchers can find out how consumers' operations are digitally recorded, analysed and used in the designs of OFPS.

### *ToP and Sociotechnical Assemblages*

The second theoretical thread from which sociotechnical assemblage views take inspiration is ToP. Whereas ANT has its strength in identifying the active/agential role of nonhuman actors and proposing an emergent, dynamic and flexible view on social phenomena, ToP endeavours to explain how everyday routinised activities come to be what they are in contemporary society. Thus, since the milestone of Warde's article *Consumption and Theories of Practice* (2005), ToP has been broadly used in consumption research, through which consumption is interpreted as a moment in almost every practice. Based on this paradigm shift, more and more consumption research looks into everyday practices including household energy use (Shove & Walker, 2014), showering or bathing (Hand et al., 2005), doing laundry (Mylan & Southerton, 2018), managing household food waste (Evans, 2014) and eating (Warde, 2016), instead of merely focusing on the process of transactions and shopping behaviour (Warde, 2014). These studies are sociotechnical in the sense that they not only present the change in everyday activities, but also involve the coevolution and embeddedness of new technologies, including central heating systems, hot water supply, washing machines and fridges. Compared with ANT, ToP posits some stability in social phenomena and agrees that institutional knowledge and structure can lie 'out there' without practices being performed at every moment. However, ToP also argues that stability is dynamic and changeable. To understand this state of 'being dynamically stable', it will be useful to explore the concept of 'practice' and how 'practices' can be treated as 'entities' and 'performances' from different perspectives.

ToP, in their multiple sense, refer to a broad range of theories which consider practice as the unit of social analysis. Practice, according to Schatzki (1996), is a nexus of doings and sayings. By doings and sayings,

Schatzki (1996) means that practice comprises both practical activities and their representations. He proposed two ways to conceptualise practice: by taking 'practice as an entity' (focusing on how doings and sayings are coordinated) and 'practice as performance' (focusing on how doings and sayings are being performed).

Firstly, when practice is considered as a coordinated entity, a related and often-cited conceptualisation is Reckwitz's definition of 'Praktik' (2002, p. 249), a concept that he defines as:

> a routinized type of behaviour which consists of several elements, interconnected to one other: forms of bodily activities, forms of mental activities, 'things' and their use, a background knowledge in the form of understanding, know-how, states of emotion and motivational knowledge.

This conceptualisation of practice has pointed out that practices are coordinated by multiple heterogeneous elements and can never be reduced to any single element (Reckwitz, 2002, p. 250). To simplify the categories of elements, Shove et al. (2012, p. 24) propose a three-element framework that has been widely used, which includes materials (objects, infrastructures, tools, hardware and the body), competence (understanding and practical knowledgeability) and meanings (emotions and motivations). Table 1 exemplifies this framework with OFPS-based food practices.

Secondly, practice can be conceptualised as performance, which is about the carrying out or the enactment of practice within specific real-life contexts (Schatzki, 1996). It is through the repetitive and recursive performance of practice that the links among constituent elements achieve a 'dynamically stable' state, and specific practices emerge, evolve or dissolve.

**Table 1** Elements of OFPS-based food provisioning practices

| *Categories of Elements* | *Explanation* | *Examples* |
|---|---|---|
| Materials | Objects, infrastructures, tools, hardware and the body. | Food, mobile phone or computer, vehicle, Internet, kitchen facilities |
| Competence | Understanding, practical knowledgeability. | Skills (of using devices, accessing the Internet), Understanding (of good food, comparing prices) |
| Meanings | Social and symbolic significance of participation. | Caring for the family, supporting local business |

A related example is the emerging practices of OFPS-based food consumption. Food consumption practices have been performed for a long time, through the embodied behaviour of shopping for food, preparing meals, cooking and eating. On the threshold of this century, with the development of mobile Internet and digital technologies, there is potential to build and reconfigure the connections among digital platforms, food producers, retailers and consumers. However, it is through consumers' repetitive use of digital platforms to access food, and retailers' digitalisation of market devices, that those reconfigurations and coordination can potentially become consolidated and normalised, and OFPS-based food practices can become everyday practices for a certain group of consumers.

The conceptualisations of practice as entity and performance address the neglect of convention, custom, habitual activities, shared knowledge and meanings in ANT, thus ToP maintains a dynamically stable view on the social level, a view that is useful in understanding the everyday embeddedness of OFPS. On one hand, when practice is considered as an entity, constituent elements such as meanings and competencies tend to play important roles in the emerging OFPS-based food consumption. For instance, the incorporation of box schemes into everyday cooking cannot happen without consumers' growing interest in sustainable and healthy eating, needless to say, the support from their cooking skills and food-matching knowledge (Fuentes & Samsioe, 2021). On the other hand, practice as performance highlights the enactment and instantiation of practice, which is guided and organised by the above meanings, conventions and practical knowledge. In this sense, ToP not only achieves a state of being dynamically stable, but also connects individuals with the social. When people perform practices, they coordinate heterogeneous elements based on specific understandings and rules. Meanwhile, through ongoing and recurring performances, elements and the ways they are coordinated are consistently reproduced and reconfigured. This process reflects that stability and transition are the results of the continuous and mutual process of coordination and performance.

### *MLP, Transition and Sociotechnical Assemblages*

The third theoretical thread contributing to sociotechnical assemblage views is MLP, which was originally proposed by Rip and Kemp (1998) and was further developed by Frank Geels (2002, 2004, 2007). MLP's contribution mainly resides in its focus on the transitions of sociotechnical regimes and how transitions happen among the interactions of entities from three different levels: sociotechnical landscapes, sociotechnical regimes and niches (Geels, 2002). Compared with ANT and ToP, it takes a vertical epistemology towards technology and society, an epistemology that puts social factors in the macro layer and considers them as exogenous to the micro-level technological innovation. Figure 1 shows how MLP can be used to understand the potential transition towards OFPS-based food consumption.

In this diagram, levels are analytical concepts to understand the dynamics of sociotechnical change and the degree of stability (Geels, 2011). Through MLP, the emergence and prevalence of OFPS can be analysed at three levels: (1) the exogenous macro-level 'sociotechnical landscapes' which highlight the technical, physical and material backdrops that affect and sustain food provisioning, such as a pandemic shock and the development of digital technologies, including improved networks, mobile devices and digital infrastructure (Dulsrud & Bygstad, 2022); (2) the stabilised, locked-in meso-level sociotechnical regimes which incorporate alignments of activities and the 'deep structure' or grammar of sociotechnical systems (for example, how consumers carry out everyday food sourcing and eating

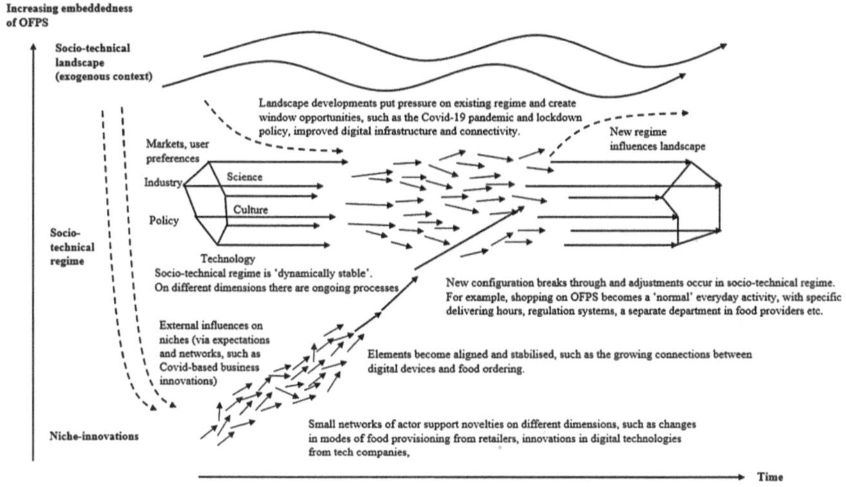


**Fig. 1** Transitioning towards OFPS-based food consumption (adapted from Schot & Geels, 2008)

in ways that are unintentionally shared by certain groups of people); (3) the micro-level niches, where radical innovations are incubated and emerge, such as the innovation of online food shopping conducted by existing food retail companies.

Each level incorporates a heterogeneous configuration of elements (termed 'dimensions' in Geels, 2004), including systems (tangible elements such as artefacts and material conditions), actors (human actors, organisations and social groups) and rules-institutions. And it is the dynamics and interactions among these elements that promote the transition of sociotechnical systems (Geels, 2011). For example, rules at the niche level are relatively loose and less crystallised (Kemp et al., 1998), protecting and encouraging actors to conduct multiple innovations in the food industry (such as the evolving algorithms for order allocations, personalised recommendations and optimised logistical systems). Meanwhile, pressures from exogenous sociotechnical landscapes continuously target the incumbent regimes, which may create windows of opportunity for emerging innovation to make a breakthrough. This process could be well exemplified by how the COVID-19 pandemic has created opportunities for online food provisioning: shortages of essential workers and closed restaurants have disrupted the existing food provisioning system. In addition, the previous perception of food consumption has also changed: shopping in stores and eating in were deemed as risky activities. As innovations on food provisioning platforms accumulated and scaled up, and landscape pressures grew heavier, existing food provisioning regimes became less stable, bringing up opportunities for online food provisioning to step in. However, since the brick-and-mortar stores are deeply entrenched in the food provisioning system, a new blended food provisioning regime emerged and became stabilised, with set rules and interdependent networks among actors (Geels, 2004).

MLP has brought in the concept of 'sociotechnical' in many senses. Firstly, it highlights the coevolution of technology and society, instead of treating them as discrete or separate or having a causal relationship such as technology determinism. For example, innovations of platform technologies and their increasing adoptions in food provision and consumption systems are consistent with broader transitions in other sociotechnical regimes towards a 'platform society' (Van Dijck et al., 2018). Thus, the understanding of emerging OFPS cannot be isolated from the broader context of platform-society transformation. Secondly, among the coevolution of digital technologies and platform society, MLP has identified heterogeneous actors or elements at different sociotechnical levels. For instance, food producers and platform companies create innovative modes of food provision (niches), consumers embrace novel attitudes and understanding of food consumption (regimes), and related policies and regulations from governments accordingly arise (landscapes). This theoretical framework combines actors from both the production side and the consumption side in empirical research (McMeekin & Southerton, 2012). Thirdly, MLP proposes a clear conceptual framework to understand how transition may happen, become stabilised and persist, a framework that is useful in conceptualising the emerging platform technologies in food systems, in terms of what exogenous social factors are creating opportunities for potential transitions, and whether platform innovations can be admitted into the incumbent regime levels (Cohen, 2012).

## 3 Sociotechnical Assemblage Views on OFPS

In the last section, I have introduced the concept of sociotechnical assemblage and explained how three theoretical threads inspire sociotechnical assemblage views. In general, ANT has highlighted the importance of nonhuman actors and how they function with human actors in the establishment of actor networks. ToP stress the investigation into everyday routinised activities and how these activities are stabilised through continuous reconfiguration of heterogeneous elements. MLP proposes a three-level paradigm to understand the emergence of sociotechnical transition. Despite their difference in ontologies ('vertical' versus 'horizontal') and in understanding of 'the social', all three theories avoid linear, one-directional causal relations between technology and society. Their commonalities can be considered as the 'golden rules' for sociotechnical assemblage thinking:

1. Social phenomena should be explored from the constituents of heterogeneous elements, including both human and nonhuman actors.
2. These human and nonhuman actors connect in specific ways to provide the settings where social phenomena take place.
3. Change and stability always coexist, and transition takes place in a dynamically stable state.

Based on these rules, I argue that OFPS should be conceptualised with sociotechnical assemblage thinking by exploring OFPS on two scales: (1) OFPS as actors which constitute sociotechnical assemblages and (2) OFPS themselves as sociotechnical assemblages (see Fig. 2).

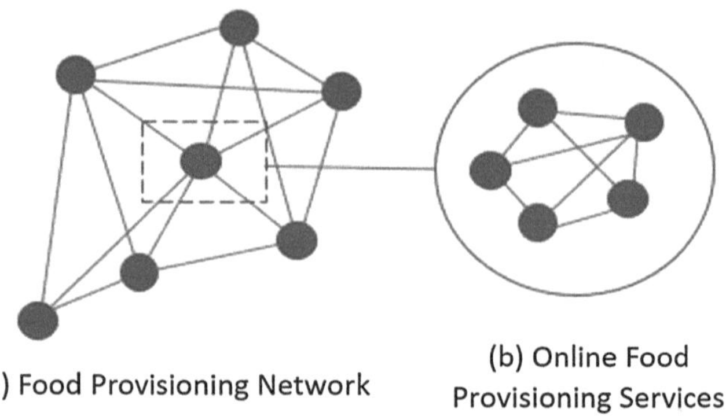


**Fig. 2** Sociotechnical assemblage thinking on OFPS

From the first angle (a), OFPS are seen as an important intermediary actor in food provision systems, which connects heterogeneous actors such as food providers, producers, consumers, retailers, food, deliverymen, transporting vehicles and so on. This intermediary role can be explained by a key feature of OFPS: multi-sidedness, which refers to their capability to connect multiple actors with multi-interfaces simultaneously. Through multi-sided digital devices, actors can share information and communicate with each other in real time (Hagiu & Wright, 2015). In practice, they not only help to match supply with potential demand, facilitating transactions among different actors, but also offer spaces for different actors to cooperate and communicate online (Van Dijck, 2013; Varian, 2010). Delivery platforms and online supermarket shopping illustrate this point clearly. The former allows consumers to order food from different restaurants within a certain geographical area, while the latter also connects consumers, delivery drivers, food and provisioners (though fewer food providers are included).

It should also be noted that the intermediary role of OFPS can take effect only when connections among actors are built. And in this way, the whole assemblage is aiming for the same goal, which is also why an assemblage cannot be reduced to discrete actors when being studied. Consequently, OFPS are both the processes of arranging multiple heterogeneous actors and the results of such arrangements, which will evolve with the whole sociotechnical network as relations among different actors reweave and are reassembled. Latour (2005, p. 208) has also mentioned these provisional and emerging features of information technology:

> information technology [...] was supposed to be an all-or-nothing affair, it has now clearly become the provisional result of a whole assemblage of plug-ins coming from completely different loci. To be a realistic whole is not an undisputed starting point but the provisional achievement of a composite assemblage.

The second angle, which may open up a more theoretical and empirical discussion, is to consider OFPS themselves as sociotechnical assemblages (b). This angle is also related to the first angle in the sense that OFPS, as 'micro' or 'sub' networks, are 'punctualised' (being taken as one single actor) or 'black-boxed' when being integrated into food systems. From this angle, OFPS are viewed as assemblages instead of simple technologies. They can function only with the contributions and linkages among their constituent actors, including intangible components such as data collection programs, algorithms, infrastructural support such as Internet services and connecting devices, and human input of all sorts of raw data. Existing research on digital platforms has already exposed these black boxes in understanding platform capitalism (Langley & Leyshon, 2017; Srnicek, 2017) and how surveillance, as well as data extraction, happens when people are incessantly connected to the Internet (Zuboff, 2019). For example, when consumers start using food provisioning platforms to access food, their personal data, purchase behaviour and preferences will all be recorded, a record which will then be used to train the platform algorithms and make recommendations of greater accuracy. The better-customised services are, the more consumers will be attracted, which in turn generates more data to help optimise the algorithms. It is this network effect logic that stimulates and enables food platforms to grow in scale and scope (Srnicek, 2017, p. 43). That is because, on one hand, users are located in complicated social networks, and their connections with other people and organisations can bring more potential users to platforms. On the other hand, this virtuous cycle will also attract venture capital, whose investment helps digital platforms to scale up (Parker et al., 2016). However, research on OFPS has not yet paid sufficient attention to what is happening behind digital technologies and their operation logic. Opening up the black box of OFPS will allow researchers to better incorporate the production side of food systems and understand the role of markets, technologies and business strategies in food provision and consumption.

## 4 Understanding OFPS-Based Food Consumption with ToP, MLP and ANT

The above sections have explained how three theoretical threads, namely, ANT, ToP and MLP, contribute to sociotechnical assemblage thinking and how OFPS and OFPS-based food consumption can be understood with sociotechnical assemblage views. Although there are some ontological differences among these theories (for example, ANT has a 'flat ontology', whereas MLP has a structured and 'vertical' one), I will borrow Warde's (2014, p. 280) argument that 'theories are instruments of selective attention', and these three theories have their own strengths and weaknesses in answering different research questions and queries. In this section, they will be used to approach two key themes in digitalised and sustainable consumption: (1) actor interactions and agencies; (2) stability and transition. The aim is to find out how different theories provoke different research queries in understanding OFPS-based food consumption.

### *Actor Interactions and Agencies*

Although ToP, MLP and ANT weigh actors and their agencies differently, all of them have included the activities of actors and interactions in their respective theoretical frameworks. ANT takes the most radical perspective that all actors, humans and nonhumans, have their agencies, which are symmetric and equal. That means when understanding OFPS-based food consumption, ANT also focuses on the agential power of nonhuman actors such as food, online food provisioning conduits and other infrastructural technologies. This focus allows researchers to open the black box of digital technologies, to see what constitutes and what is behind OFPS. For example, what is the logic behind different market devices on OFPS, such as e-trolley, filtering functions, minimum spending and box schemes? Moreover, in terms of actor interactions, ANT is aligned with symmetric agencies in the sense that it explores interactions among all actors, be they human-to-human, human-to-nonhuman or nonhuman-to-nonhuman. For instance, the concepts of scripting and domestication illustrate how digital technologies shape consumers' food shopping activities and vice versa. Despite its advantages in equal agencies and a focus on connections, ANT somehow takes 'a potentially leaden view of stuff' because of neglecting the ordinary questions about what nonhuman actors are actually doing (Shove et al., 2012). Also, because all objects and human beings in the world are to some extent connected to other objects and human beings, ANT may extend the network infinitely and fail to identify key links worthy of more attention.

ToP maintains that agencies are distributed among human and nonhuman actors (Sahakian & Wilhite, 2014), but also maintains that not all actors (and their agencies) are equivalent or equal, a stance which can be considered as somewhere between that of MLP and ANT. On one hand, nonhuman actors, such as food, infrastructure and household facilities, are considered as essential elements in practices. All elements of practice have the capability to be the source of a particular performance of practice to some extent, which means a change in any element may imply more than the change in itself, but also the way different elements constitute practices. For example, improved cooking skills may transform one's eating habits, and the emergence of OFPS has introduced new shopping routines. On the other hand, not all actors are equal. People, as the carriers of practices, actively coordinate heterogeneous elements based on their shared understanding and practical rules to fulfil specific goals. OFPS-based food practices can be coordinated only based on consumers' practical knowledge, and they can be performed only via consumers' operation of applications or websites, also via the delivery action from couriers. Elements with more power may also affect complexes of other elements, such as policy and digital technologies (Shove & Walker, 2010). These differences in agential power allow researchers and policymakers to identify important places that serve as the intersection of different practices when promoting sustainable transition, such as the kitchen and the fridge (Joosse & Marshall, 2020). Another strength of ToP is its focus on the performance of everyday practices, through which researchers can find out how heterogeneous elements evolve and interact in everyday life, and practices which may be overlooked because they are embedded in normalised activities and are always taken for granted.

As for MLP, it locates actors within three conceptual levels: niche, regime and landscape. And since its purpose is to provide a theoretical framework to understand sociotechnical transitions, the actor interactions it highlights also centre around the transition process. Specifically, it explores how changes in exogenous social contexts create windows of opportunity for transition and how technological innovation at the niche level grows into stabilised regimes (Geels, 2011). Take OFPS-based food practices as an example. At the niche level, it asks how innovative actors such as R&D departments in food companies and tech companies work on radical innovation to connect multiple food producers and improve logistical services. Meanwhile, at the landscape level, it aims to find out how 'macro' actors, such as the state, market and even the pandemic destabilise the existing food provisioning regimes to allow transitions to happen. Thus, it can be argued that MLP's advantage lies in that it has provided a relatively clear conceptual framework to understand actors and their interactions in different 'levels' (Genus & Coles, 2008). However, it speaks less about them on a daily basis, such as how people use emerging technologies in their everyday life, ignoring everyday practices that may also be a source of innovation (McMeekin & Southerton, 2012; Seyfang & Smith, 2007). Also, agencies are presented in the form of 'bounded rationality (routines, search activities, trial-and-error learning) and interpretive activities' (Geels, 2011, p. 30), which are applicable only to human beings and social groups. As a result, MLP may have ignored the role of nonhuman actors and how human beings interact with them.

To sum up, it is obvious that actors, their agencies and interactions should be the key queries when studying OFPS and OFPS-based food consumption. The very first question coming to researchers should be 'What are the actors that constitute OFPS-based food practices?' Not only should the actors simply be mapped out, but what is behind those actors and how they function are also worth further investigation. This proposition echoes ANT's proposition of 'opening up the black box', an activity which is particularly important in the context of platform society and the prevalence of algorithms and big data. Then, the interactions among different actors, particularly human-technology interactions should be explored (McMeekin & Southerton, 2012, p. 357). To be more specific, researchers are encouraged to find out how OFPS shape consumer behaviour by collecting and using their data, and what marketing devices are made based on consumer data. In addition, it is also important to know how consumers are using these OFPS and technologies in their everyday life, what the roles of these OFPS are and how they are related to other everyday routines. Only in these ways, by combining the production and consumption ends, can researchers better understand how different actors in food systems are connected and integrated and how such connections become embedded in consumers' everyday life.

### *Stability and Transition*

Although MLP, ANT and ToP have different explanations for how stabilities are maintained and how transitions happen, all of them try to avoid a dichotomous view on stability and transition and take the 'dynamically stable' view instead. MLP explains transition as the result of interactions among entities within and between its three levels. It happens as the innovation in the niche level is admitted into the regime levels and becomes stabilised under the exogenous forces from the sociotechnical landscape. For example, with the development of Internet and mobile technologies, takeaway delivery service has transformed into an application or website-based service from its original phone-call version (Liu & Chen, 2021). In terms of stability, it stresses the concept of path dependencies and lock-ins, which tend to keep the existing sociotechnical regime stable (Geels, 2012). During the pandemic, many people took up online grocery shopping as a conduit to source fresh food, while shopping in-store became a less frequent practice by some. However, in the new normal (when most restrictions are lifted), people start going back to the brick-and-mortar store and online grocery shopping is less popular. But even some people return to shopping in physical stores, some still retain their online shopping practices. The reasons behind such differences are worth exploring. With the MLP framework, researchers are encouraged to find out what makes or hinders a successful transition by studying different types of path dependencies and lock-ins during this process (Geels, 2010). Also, it encourages researchers to find out how particular transitions happen. For example, during the transition towards sustainable consumption, what innovations are useful? How can these innovations get into sociotechnical regimes of everyday life? What are the roles of government, policies and other macro entities? However, MLP tends to focus more on the possibility and potentiality of transition, but it speaks less about how the new regime persists in everyday life (Shove & Walker, 2010).

ToP reckons that practice simultaneously represents forms of stability and change (McMeekin & Southerton, 2012). Transitions happen as the constituent elements of practice change, as the way they are coordinated as practices change and as how they relate to other practices change (Shove et al., 2012). Consequently, transition and stability are more about the reconfiguration and rearrangements of practice and practice elements. For example, Hand et al. (2005) delineate how the practice of showering has come to be performed with greater frequency because of changes in material elements (increasing electricity and hot water supply, improved bathroom facilities), changes in meanings (the connotation of cleanliness and hygiene) and changes in competence (the 'knowledge' that one should have a shower after physical activity). The transformation from bathing to showering may also be explained by the changing temporal scheduling of other everyday practices (for example, showering can be slotted into narrow time frames to fit a more standardised and shared schedule). Thus, in the context of sustainable transition in food consumption, researchers can look into how sustainable food practices come about by exploring the changes in constituent elements, the ways these elements are connected and the ways food practices are related to other everyday activities. As for stability, ToP believe that it is achieved through repetitive configuration and coordination of elements and the recursively reproduced and routinised performance of practices (McMeekin & Southerton, 2012; Shove et al., 2012). Based on this understanding, researchers can use ethnographic methods to study how consumers conduct food practices in everyday life. How are food purchase, storage, preparation, cooking, eating and waste management performed recursively and repetitively? And during these processes, how is tacit knowledge, such as shared understanding and skills, cultivated? Additionally, ToP are useful in explaining the 'sticky' and 'deeply embedded' everyday practices, practices that sometimes present themselves as the 'hybrid' or 'intermediate' stage towards transition. This explanation will help researchers to examine why and how such a stage occurs and why transitions sometimes take long and even transform into other formats.

As for ANT, it adopts a more processual and emergent view of stability and change. Food provisioning networks exist only as connections among multiple actors (such as digital technologies, food, consumers, food providers, couriers and vehicles) are built and performed. As a result, a food provisioning network is stable only because consumers repetitively and recurrently use digital platforms to buy food. Consequently, stability (temporary) is the emergent outcome of interactions, but not something that actors can follow and learn from. Likewise, what sociologists called 'social contexts' or 'the social' is only the outcome of actor connections instead of something lying 'out there' (Latour, 2005, pp. 34–35). With these emergent and relations perspectives, ANT encourages researchers to always describe the networks of OFPS-based food consumption first, because networks are always dynamic. Specifically, what actors constitute the networks and how are networks constituted? Then, researchers should study the interactions among actors, and how those interactions change/maintain as time goes by. Only in this way can the state of being dynamically stable be mapped out.

## 5 Conclusion

This chapter starts by introducing digital platforms and how they are becoming increasingly popular in food consumption. Although the connections between food consumption and digital technologies are receiving unprecedented attention, there are still insufficient theoretical and empirical investigations of OFPS from which people directly access food in everyday life. Given the complexity of food consumption, technological development and their co-evolvement, this chapter aims for a theoretical approach to conceptualise OFPS and their embeddedness in everyday life. Sociotechnical assemblage views, which are the theoretical stances shared by different theories aiming to explore the relationships between technology and society, are considered as an insightful approach to understanding OFPS and OFPS-based food consumption.

ANT, ToP and MLP are three key theoretical strands that constitute and substantiate sociotechnical assemblage thinking. Their commonalities are considered as the 'golden rules' for sociotechnical assemblage views: heterogeneous actors, relational views and dynamic stability. Nevertheless, all of them have their own advantages in exploring particular types of research queries. With a clear conceptual framework of sociotechnical transition, MLP is useful in identifying the key factors promoting transition and in describing the transitional process, be it successful or not. Meanwhile, it provides some explanation of unsuccessful transitions but fails to offer detailed analyses of actor agency and interactions. ANT solves this problem by involving both human and nonhuman actors in actor networks. Its advantage lies in its emphasis on actors and their interactions, an emphasis which encourages studies to explore different actors in detail, finding out what is behind each actor and how it affects actor interactions. To achieve this goal, researchers are required to 'follow the actors' and 'open the black box' in empirical data collection.

As discussed in the previous section, ToP can be considered as an intermediate approach between MLP and ANT. It not only adopts a view of distributed and unequal agencies among both human and nonhuman elements surrounding OFPS-based food consumption but also conceptualises practices as dynamically stable through analysing practice as an entity and performance. However, it should be noted that the true value and outstanding advantage of ToP are based on its priority in everyday activities that have been normalised and routinised. Based on this understanding, ToP, aligned with MLP and ANT, can provoke some potential research topics that should be prioritised in future research: (1) the identification and exploration of nonhuman actors/elements in food consumption practices; (2) actor/element interactions that underpin the performance of food consumption practices (in particular, human–nonhuman interactions); (3) the evolving trajectories of actors/elements and the ways they are connected; (4) the performance of everyday food practices on a long-term scale; (5) the transition of food consumption and how it relates to macro and micro sociotechnical change. These are key themes from a sociotechnical assemblage perspective on OFPS and OFPS-based food consumption. Most importantly, researchers should be clear about their research questions and adopt the appropriate theoretical approach, because none of the above theories are suitable for all queries in future food research.

## References

Akrich, M. (1992). The de-scription of technical objects. In W. E. Bijker & J. Law (Eds.), *Shaping technology/building society: Studies in sociotechnical change* (pp. 205–224). MIT Press.

Callon, M. (1999). Actor-network theory—The market test. *The Sociological Review, 47*(1 _suppl), 181–195. https://doi.org/10.1111/j.1467-954X.1999.tb03488.x

Cohen, M. J. (2012). The future of automobile society: A socio-technical transitions perspective. *Technology Analysis & Strategic Management, 24*(4), 377–390. https://doi.org/10.1080/09537325.2012.663962

Dannenberg, P., Fuchs, M., Riedler, T., & Wiedemann, C. (2020). Digital transition by COVID-19 pandemic? The German food online retail. *Tijdschrift voor Economische en Sociale Geografie, 111*(3), 543–560. https://doi.org/10.1111/tesg.12453

de Reuver, M., Sørensen, C., & Basole, R. C. (2018). The digital platform: A research agenda. *Journal of Information Technology, 33*(2), 124–135. https://doi.org/10.1057/s41265-016-0033-3

Deleuze, G., & Parnet, C. (1987). *Dialogues.* Columbia University Press.

Douglas, D. G. (2012). *The social construction of technological systems: New directions in the sociology and history of technology*. MIT Press.

Dulsrud, A., & Bygstad, B. (2022). Digital ecosystems as social institutions: Exploring the role of consumption through four research streams of digital ecosystems. *Consumption and Society, 1*(1), 99–119. https://doi.org/10.1332/QUPH6141

Evans, D. M. (2014). *Food waste: Home consumption, material culture and everyday life*. Bloomsbury Publishing.

Evans, D. M., Jackson, P., Truninger, M., & Baptista, J. A. (2022). The ontological politics of freshness: Qualities of food and sustainability governance. *Environment and Planning A: Economy and Space, 54*(3), 461–476. https://doi.org/10.1177/0308518X211059834

Farías, I., & Bender, T. (2012). *Urban assemblages: How actor-network theory changes urban studies*. Routledge.

Fuentes, C., & Samsioe, E. (2021). Devising food consumption: Complex households and the socio-material work of meal box schemes. *Consumption Markets & Culture, 24*(5), 492–511. https://doi.org/10.1080/10253866.2020.1810027

Fuentes, C., Samsioe, E., Fuentes, M., & Ivre, J. (2024). Making the consumption of food circular: The Karma app and the re-qualifications of surplus food. In *Digital food provisioning in times of multiple crises: How social and technological innovations shape everyday consumption practices*. Palgrave Macmillan.

Fuentes, C., & Sörum, N. (2019). Agencing ethical consumers: Smartphone apps and the socio-material reconfiguration of everyday life. *Consumption Markets & Culture, 22*(2), 131–156. https://doi.org/10.1080/10253866.2018.145642

Geels, F. W. (2002). Technological transitions as evolutionary reconfiguration processes: A multi-level perspective and a case-study. *Research Policy, 31*(8–9), 1257–1274. https://doi.org/10.1016/S0048-7333(02)00062-8

Geels, F. W. (2004). From sectoral systems of innovation to socio-technical systems: Insights about dynamics and change from sociology and institutional theory. *Research Policy, 33*(6–7), 897–920. https://doi.org/10.1016/j.respol.2004.01.015

Geels, F. W. (2007). Transformations of large technical systems: A multilevel analysis of the Dutch highway system (1950–2000). *Science, Technology, & Human Values, 32*(2), 123–149. https://doi.org/10.1177/0162243906293883

Geels, F. W. (2010). Ontologies, socio-technical transitions (to sustainability), and the multi-level perspective. *Research Policy, 39*(4), 495–510. https://doi.org/10.1016/j.respol.2010.01.022

Geels, F. W. (2011). The multi-level perspective on sustainability transitions: Responses to seven criticisms. *Environmental Innovation and Societal Transitions, 1*(1), 24–40. https://doi.org/10.1016/j.eist.2011.02.002

Geels, F. W. (2012). A socio-technical analysis of low-carbon transitions: Introducing the multi-level perspective into transport studies. *Journal of Transport Geography, 24*, 471–482. https://doi.org/10.1016/j.jtrangeo.2012.01.021

Genus, A., & Coles, A. M. (2008). Rethinking the multi-level perspective of technological transitions. *Research Policy, 37*(9), 1436–1445. https://doi.org/10.1016/j.respol.2008.05.006

Gilliland, J., Sadler, R., Clark, A., O'Connor, C., Milczarek, M., & Doherty, S. (2015). Using a smartphone application to promote healthy dietary behaviours and local food consumption. *BioMed Research International, 2015*, 841368. https://doi.org/10.1155/2015/841368

Hagiu, A., & Wright, J. (2015). Multi-sided platforms. *International Journal of Industrial Organization, 43*, 162–174. https://doi.org/10.1016/j.ijindorg.2015.03.003

Hand, M., Shove, E., & Southerton, D. (2005). Explaining showering: A discussion of the material, conventional, and temporal dimensions of practice. *Sociological Research Online, 10*(2), 101–113. https://doi.org/10.5153/sro.1100

Hassen, T. B., El Bilali, H., Allahyari, M. S., Berjan, S., & Fotina, O. (2021). Food purchase and eating behavior during the COVID-19 pandemic: A cross-sectional survey of Russian adults. *Appetite, 165*, 105309. https://doi.org/10.1016/j.appet.2021.105309

Heidenstrøm, N., & Hebrok, M. (2022). Towards realizing the sustainability potential within digital food provisioning platforms: The case of meal box schemes and online grocery shopping in Norway. *Sustainable Production and Consumption, 29*, 831–850. https://doi.org/10.1016/j.spc.2021.06.030

Joosse, S., & Marshall, M. (2020). Fridge stories and other tales from the kitchen: A methodological toolbox for getting closer to everyday food practices. *Food, Culture & Society, 23*(5), 608–626. https://doi.org/10.1080/15528014.2020.1778921

Kemp, R., Schot, J., & Hoogma, R. (1998). Regime shifts to sustainability through processes of niche formation: The approach of strategic niche management. *Technology Analysis & Strategic Management, 10*(2), 175–198. https://doi.org/10.1080/09537329808524310

Langley, P., & Leyshon, A. (2017). Platform capitalism: The intermediation and capitalisation of digital economic circulation. *Finance and Society, 3*(1), 11–31. https://doi.org/10.2218/finsoc.v3i1.1936

Latour, B. (2005). *Reassembling the social: An introduction to actor-network-theory.* Oxford University Press.

Latour, B., Jensen, P., Venturini, T., Grauwin, S., & Boullier, D. (2012). 'The whole is always smaller than its parts'—A digital test of Gabriel Tarde's monads.

*The British Journal of Sociology, 63*(4), 590–615. https://doi.org/10.1111/j.1468-4446.2012.01428.x

Latour, B., & Johnson, J. (1988). Mixing humans with non-humans? Sociology of a few mundane artefacts. *Social Problems, 35*, 298–310. https://doi.org/10.2307/800624

Liu, C., & Chen, J. (2021). Consuming takeaway food: Convenience, waste and Chinese young people's urban lifestyle. *Journal of Consumer Culture, 21*(4), 848–866.

Lupton, D. (2021). Young people's use of digital health technologies in the global north: Narrative review. *Journal of Medical Internet Research, 23*(1), e18286. https://doi.org/10.2196/18286

McMeekin, A., & Southerton, D. (2012). Sustainability transitions and final consumption: Practices and socio-technical systems. *Technology Analysis & Strategic Management, 24*(4), 345–361. https://doi.org/10.1080/09537325.2012.663960

Müller, M. (2015). Assemblages and actor-networks: Rethinking socio-material power, politics and space. *Geography Compass, 9*(1), 27–41. https://doi.org/10.1111/gec3.12192

Müller, M., & Schurr, C. (2016). Assemblage thinking and actor-network theory: Conjunctions, disjunctions, cross-fertilisations. *Transactions of the Institute of British Geographers, 41*(3), 217–229. https://doi.org/10.1111/tran.12117

Mylan, J., & Southerton, D. (2018). The social ordering of an everyday practice: The case of laundry. *Sociology, 52*(6), 1134–1151. https://doi.org/10.1177/0038038517722932

Oncini, F. (2024). Online food providers and where to find them: Pipelines, platforms, and the rise of dark stores. In *Digital food provisioning in times of multiple crises: How social and technological innovations shape everyday consumption practices*. Palgrave Macmillan.

Parker, G. G., Van Alstyne, M. W., & Choudary, S. P. (2016). *Platform revolution: How networked markets are transforming the economy and how to make them work for you*. W. W. Norton & Company.

Reckwitz, A. (2002). Toward a theory of social practices: A development in culturalist theorizing. *European Journal of Social Theory, 5*(2), 243–263. https://doi.org/10.1177/13684310222225432

Rip, A., & Kemp, R. (1998). Technological change. In S. Rayner & E. L. Malone (Eds.), *Human choice and climate change: Vol. II, resources and technology* (pp. 327–399). Battelle Press.

Sahakian, M., & Wilhite, H. (2014). Making practice theory practicable: Towards more sustainable forms of consumption. *Journal of Consumer Culture, 14*(1), 25–44. https://doi.org/10.1177/1469540513505607

Samsioe, E., & Fuentes, C. (2022). Digitalizing shopping routines: Re-organizing household practices to enable sustainable food provisioning. *Sustainable*

*Production and Consumption, 29*, 807–819. https://doi.org/10.1016/j.spc.2021.07.019

Sawyer, S., & Jarrahi, M. H. (2014). Sociotechnical approaches to the study of information systems. In *Computing handbook: Information systems and information technology* (3rd ed., pp. 5-1–5-27). CRC Press. https://doi.org/10.1201/b16768

Schatzki, T. R. (1996). *Social practices: A Wittgensteinian approach to human activity and the social.* Cambridge University Press.

Schneider, T., Eli, K., McLennan, A., Dolan, C., Lezaun, J., & Ulijaszek, S. (2019). Governance by campaign: The co-constitution of food issues, publics and expertise through new information and communication technologies. *Information, Communication & Society, 22*(2), 172–192. https://doi.org/10.1080/1369118X.2017.1363264

Schot, J., & Geels, F. W. (2008). Strategic niche management and sustainable innovation journeys: Theory, findings, research agenda, and policy. *Technology Analysis & Strategic Management, 20*(5), 537–554. https://doi.org/10.1080/09537320802292651

Seyfang, G., & Smith, A. (2007). Grassroots innovations for sustainable development: Towards a new research and policy agenda. *Environmental Politics, 16*(4), 584–603. https://doi.org/10.1080/09644010701419121

Shove, E., Pantzar, M., & Watson, M. (2012). *The dynamics of social practice: Everyday life and how it changes.* Sage.

Shove, E., & Walker, G. (2010). Governing transitions in the sustainability of everyday life. *Research Policy, 39*(4), 471–476. https://doi.org/10.1016/j.respol.2010.01.019

Shove, E., & Walker, G. (2014). What is energy for? Social practice and energy demand. *Theory, Culture & Society, 31*(5), 41–58. https://doi.org/10.1177/0263276414536746

Srnicek, N. (2017). *Platform capitalism.* Polity Press.

Van Dijck, J. (2013). 'You have one identity': Performing the self on Facebook and LinkedIn. *Media, Culture & Society, 35*(2), 199–215. https://doi.org/10.1177/0163443712468605

Van Dijck, J., Poell, T., & De Waal, M. (2018). The platform society: Public values in a connective world. *Oxford University Press.* https://doi.org/10.1093/oso/9780190889760.001.0001

Van Droogenbroeck, E., & Van Hove, L. (2017). Adoption of online grocery shopping: Personal or household characteristics? *Journal of Internet Commerce, 16*(3), 255–286. https://doi.org/10.1080/15332861.2017.1317149

Varian, H. R. (2010). Computer mediated transactions. *American Economic Review, 100*(2), 1–10. https://doi.org/10.1257/aer.100.2.1

Venturini, T., Jacomy, M., Meunier, A., & Latour, B. (2017). An unexpected journey: A few lessons from sciences Po médialab's experience. *Big Data & Society, 4*(2). https://doi.org/10.1177/2053951717720949

Warde, A. (2005). Consumption and theories of practice. *Journal of Consumer Culture, 5*(2), 131–153. https://doi.org/10.1177/1469540505053090

Warde, A. (2014). After taste: Culture, consumption and theories of practice. *Journal of Consumer Culture, 14*(3), 279–303. https://doi.org/10.1177/1469540505053090

Warde, A. (2016). *The practice of eating*. Polity Press.

Zuboff, S. (2015). Big other: Surveillance capitalism and the prospects of an information civilization. *Journal of Information Technology, 30*(1), 75–89. https://doi.org/10.1057/jit.2015.5

Zuboff, S. (2019). *The age of surveillance capitalism: The fight for a human future at the new frontier of power*. Profile Books.

# Appendices

This methodological appendix reports the three data collection techniques used during the PLATEFORMS study. This additional section aims to enable and encourage the possibility of updating, replicating, and validating the results presented in this book. As an aid to readers, Appendix A presents the *codebook* and *codesheet* we developed and used for the systematic mapping of online food platforms. Appendix B is the interview guideline used to conduct the household study. Finally, Appendix C is the online questionnaire utilized to explore consumer preferences for online food platforms' value propositions and the effect of COVID-19 on platforms' customer base.



A. Dulsrud, F. Forno (eds.), *Digital Food Provisioning in Times of Multiple Crises*, Consumption and Public Life,
https://doi.org/10.1007/978-3-031-46323-5

# Appendix A: Systematic Mapping of Online Food Platforms

## (A) Codebook [Reference Manual to Ensure Consistency and Reliability in the Coding Process]

*1. Classification of Pipelines and Platforms*

The distinction between platform and pipelines is of utmost importance for both theoretical and empirical reasons. It is important to clarify how to distinguish between the two in order to achieve consistency in the Plateform database.

On the basis of the available literature and the practical experience acquired so far we now suggest the following definition (from Parket et al., 2016) and guidelines to recognize the difference between a platform and a pipeline. We hope this might clear things up and help everyone select the appropriate response category.

PIPELINE: "A business that employs a step-by-step arrangement for creating and transferring value, with producers at one end and consumers at the other. A firm first design a product or service. Then the product is manufactured and offered for sale, or a system is put in place to deliver the service. Finally, a customer shows up and purchases the product or service." P.6

1. Pipeline: Single-sided market and traditional e-commerce. This is usually the most prevalent form of online food shopping. The



A. Dulsrud, F. Forno (eds.), *Digital Food Provisioning in Times of Multiple Crises*, Consumption and Public Life,
https://doi.org/10.1007/978-3-031-46323-5

provisioning service selects food products which are sold through the website to consumers. Producers do not get in contact with the consumers nor rely in the provisional service as an intermediary. They may or may not be presented in the website.

In fact, the online order and the shipment are exclusive responsibility of the provisioning service. Pipelines do not allow producers to register to the website as producers, although it might be possible to find in the website a digital form (e.g. a contact link) that allows producers to signal themselves to the provisioning service.

PLATEFORM: "A business based on enabling value creating interactions between external producers and consumers. The platform provides an open, participative infrastructure for these interactions and sets governance conditions for them. The platform's overarching purpose: to consummate matches among users and facilitate the exchange of goods, services, or social currency, thereby enabling value creation for all participants." P.5

2. Platform: two (or multi) sided market. Platforms act almost exclusively as intermediary between producers and consumers (B2C), consumers and consumers (C2C), producers and producers (B2B). The primary role of platforms is to facilitate the match between the two parts, by offering a web-space that allows the contact or the transaction between the two parts. The platform offers the two parts a tool (the platform) to get in contact and does not directly acquire products from the producers. However, the platform may be involved in the shipping and might actively select which producers "fit" the platform mission. A platform can be recognized and distinguished from a pipeline by the presence of at least one of the following characteristics:

(a) The website has two separate sections: one dedicated to producers and one dedicated to consumers (or other businesses, in the case of B2B platforms) where they can register in order to "start selling" or "start buying" food products.
(b) The website clearly states that the online orders are directly forwarded to the producers who have the responsibility to manage them.
(c) The website offers an online window to producers (usually in exchange of a monthly/annual fee) that are thus allowed to use the platform to sell their products to the consumers.
(d) The website offers to consumers (C2C) a service to get in contact in order to exchange or sell food products to each other

*2. Distinction Between Brand, Category, and Product*

1. Brand: An image that defines a product, or an array of products, which are made or packed by one producer. Example: Pasta Mancini: https://shop.pastamancini.com/. The producer makes the pasta, brands it, and sells it via his website.

2. Category: A food category groups together food products that fall under the same conceptual umbrella. Since categories can change depending on the level of specification, it is important to define the food categories we are going to consider as worth of our attention. Thus far, our food categories are the following ones (new ones have been introduced, some have been unpacked):

- Bakery (both fresh and packed, sweet and salty)
- Cereals (pasta, wheat, flour, rice, etc.)
- Meat (fresh)
- Fish (fresh)
- Fruit (fresh)
- Vegetables (fresh)
- Alcoholic drinks
- Non-alcoholic drinks
- Preserves (in can, sugar, vinegar, oil, salt, or fat)
- Snacks/confectionary
- Cured meat
- Dairy products
- Ready-made meals

3. Product: A food product is a single identifiable item. Therefore we think of the cured meat category as containing several cured meat products (e.g. pork sausages, reindeer ham, jamon serrano), or the dairy category as containing several milk-based products (e.g. sheep's milk cheese, yogurt, cow milk). Depending on the provisioning service, products may be purchased one by one or in a bundle (as, for instance, in the case of provisioning services selling vegetables and fruit box schemes).

3. Exclusion and Inclusion Criteria

*We will exclude*

1. Online provisioning services selling food products of one brand only or from one producer only. You can recognize these websites as *all* products sold have the brand/logo on top of their products or the website is owned by a single producer.

2. Online provisioning services that deliver cooked meals from restaurants.

*We will include*

All pipelines and platforms selling at least two product categories and C2C provisioning services.

4. Search Engine

Using Google as a search engine guarantees a deeper reach, as the algorithm can capture information that other engines cannot. Since we are interested to reach an analytical sample as close as possible to the actual population, Google guarantees better results. However, it is necessary to repeat the search with another service that does not track users in order to validate the first search we have made. Repeat the search using the service called Duckduckgo.com with the same keywords (see Sect. 5). Remember to select your country before typing the keywords.

Just to be sure, please perform an additional search on Google and Apple APPs marketplaces with the same keywords, as some platforms might be present just as native applications and not appear on Google. This will not take long (in Italy we found just three), but it might increase the number of platforms and suggest new partners for WP2—if this happens, please analyze all sections with the revised codebook.

5. Keywords

The following keywords will be used in the search. Feel free to adapt them in your own language as you please.

Online grocery shopping

1. Online food shopping
2. Grocery shopping online
3. Box scheme
4. Online food
5. Organic food online shopping
6. Food apps

Run the analysis the first time and then add next to each keyword also the local areas (possibly NUTS2, in order to refine the search at the local level (e.g. online shopping Trentino)).

## (B) Codesheet [Guidelines for Coding and Response Categories]

| *Q1* | *Country* |
|---|---|

Description: Country of the provisioning service
Guidelines for the coding: Select the country from the dropdown menu
Main issues: In very few cases (e.g. Amazon pantry), there might be multiple countries for a single provisioning service
Solution: Select your country
Response categories:

1. Italy
2. Norway
3. Sweden
4. Germany
5. Ireland

| *Q2* | *Id* |
|---|---|

Description: Id of the provisioning service
Guidelines for the coding: Add progressive number
Open response

| *Q3* | *Name* |
|---|---|

Description: Name of the provisioning service
Guidelines for the coding: Add full name of the provisioning service
Open response

| *Q4* | *Website link* |
|---|---|

Description: Link of the provisioning service website
Guidelines for the coding: Add the website link
Open response

| *Q5* | *Pipeline or Platform* |
|---|---|

Description: (from Parker et al., 2016) "Platform revolution"
Guidance for the coding: see the Codebook
Response categories:

1. Pipeline
2. Platform

| *Q6* | *Short description of the provisioning service* |
|---|---|

Description: Short description of the provisioning service

Guidelines for the coding: Add a short description of the provisioning service functioning in English

Open response

| *Q7* | *Type of provisioning service* |
|---|---|

Description: Type of sale system of the provisioning service

Guidelines for the coding: Select between the following categories:

Response categories:

1. Mass retail channel (general, specialized, discount)
2. Independent store (it needs to have a physical store—maximum of three stores)
3. Online only (both mono and multi)
4. Alternative food network/direct sales (there needs to be reference to networks such as solidarity purchasing groups or small associations of customers)

| *Q8* | *Box scheme* |
|---|---|

Description: The provisioning service has a plan that allows customers to receive groceries on a regular basis.

Guidelines for the coding: Select "yes" only if this option is available on the website and it is clearly stated. To be a box scheme, the user must have the possibility to purchase a subscription that allows him to regularly (e.g. once a week; once a month) receive the food without having to order products from the website every time. Select "no" of this option is not available.

Response categories:

1. Yes
2. No

| *Q9* | *Prevalent form of commercial transaction* |
|---|---|

Description: Type of commercial transaction between the provisioning service and the client. B2B refers to transactions between the provisioning service and other firms; B2C between the provisioning service and consumers; C2C between consumers.

Guidelines for the coding: Distinguish between the prevalent form of transaction.

Response categories:

1. B2B
2. B2C
3. C2C
4. B2B & B2C (select this only if you can clearly identify the two commercial souls in the website)
5. Other

| *Q10* | *Year of foundation of the food provisioning service* |
|---|---|

Description: Year of foundation of the provisioning service.

Guidelines for the coding: Use the service web.archive.org or the section regarding the history of the provisioning service to retrieve the plausible year of foundation. Select "not reported" if this information is not retrievable.

Open response (4 digits) or "not reported"

| *Q11* | *Year of foundation of the current webpage as an online store* |
|---|---|

Description: Year of foundation of the provisioning service as an online provisioning service

Guidelines for the coding: Use the service web.archive.org to visit the previous version of the website so to retrieve the plausible year of foundation as an ONLINE food provisioning service. Select "not reported" if this information is not retrievable.

Open response (4 digits) or "not reported"

| *Q12* | *Location of the food provisioning service* |
|---|---|

Description: City where the provisioning service is located

Guidelines for the coding: Add the city or the region where the provisioning service is located. Some provisioning services (usually independent stores and mass retail channels) may operate in several cities: if this is the case add the region (e.g. northern Italy or Tuscany). This information can be found at the bottom of the website or in the website information page. Select "not reported" if this information is not retrievable.

Open response or "not reported"

| *Q13* | *Market geographical scope* |
| --- | --- |

Description: Reach of the products shipped by the provisioning service

Guidelines for the coding: Information are usually in the website page containing shipment information. If this is not the case, pretend to purchase food and look at the options.

Response categories:

1. Local/regional (customers can receive/pick up products only from/in a limited area)
2. National (the provisioning service ships products nationally)
3. International (the provisioning service ships products in Europe or in other continents)

| *Q14* | *Producer's geographical area* |
| --- | --- |

Description: Origin of the products sold by the provisioning service

Guidelines for the coding: Multiple responses are possible. Information can be found in the provisioning service description and by looking at the products sold.

Response categories (all that apply):

1. Local/regional (select if you can clearly identify a small/local producer or there is reference to the artisanal nature of the product, or food is sold as representing the cultural heritage of the area)
2. National (select if you can recognize famous national brands, or if you are sure that these products are manufactured in the national context)
3. International (select if you can find products that come from other countries)

| *Q15* | *Market orientation* |
|---|---|

Description: Information on how the earnings are used by the provisioning service—profit or non-profit

Guidelines for the coding: Select "non-profit" if the provisioning service clearly has an organizational form that can be linked to a non-profit orientation (e.g. association or cooperative) or when there is enough evidence that earnings are invested for organizational purposes (e.g. it is clearly stated). Select "for profit" all the other times.

Response categories:

1. Profit
2. Non-profit
3. Not reported

| *Q16* | *Type of organization* |
|---|---|

Description: The organizational form of the provisioning service

Guidelines for the coding: Select between enterprise, association, cooperative, informal group/network. Select association or cooperative when the organizational form is clearly provided in the website. Select informal group/network when the website is just a tool used by a small group of users to organize themselves but no clear organizational form is identifiable. Select enterprise all the other times. Not reported if this information is not present at all.

Response categories:

1. Enterprise
2. Association/cooperative
3. Informal group/network
4. Not reported

| *Q17* | *Number of employees* |
|---|---|

Description: Number of employees of the provisioning service

Guidelines for the coding: In the website of the provisioning service, look whether information regarding the number of employees is clearly stated. Select "not reported" if this information is not retrievable.

Open response or "not reported"

| | |
|---|---|
| *Q18* | *Bakery, cereals, meat (fresh), fish (fresh), fruit (fresh), vegetables (fresh), alcoholic drinks, non-alcoholic drinks, preserves, snacks/confectionary, cured meat, dairy products, ready-made, other* |

Description: Type of food products sold by the provisioning service

Guidelines for the coding: Look at the list of products sold by the provisioning service and select "yes" or "no." Tick YES if at least one product belonging to one category is sold. Select NO otherwise.

Response categories:

1. Yes
2. No

| | |
|---|---|
| *Q19* | *Selling mechanism* |

Description: information on the selling mechanism connecting producers, provisioning service, and consumers.

Guidelines for the coding: Select "producers sell to consumers" when the provisioning service clearly and solely operates as an intermediary connecting producers and consumers. Select "producers sell to the provisioning service" when the provisioning service sells products which are produced by others. Select "specify" and add a comment with a description when one of the two mechanisms is not clearly identifiable.

Response categories:

1. Producers sell to consumers
2. Producers sell to the provisioning service
3. Specify (please explain)

| | |
|---|---|
| *Q20* | *Pricing mechanism* |

Description: Information on how the price of the food products is set

Guidelines for the coding: Select "fixed price set by the producers" when this is clearly stated in the website. Select "fixed price set by the provision service" if this is clearly stated in the website. Select "differentiated price based on consumer features" when this is clearly stated in the website (e.g. there are options to become a premium member or prices change depending on consumers' profiles). Select not reported if the pricing mechanism is not reported.

Response categories:

1. Fixed price set by the producers
2. Fixed price set by the provisioning service
3. Differentiated price based on consumer features
4. Not reported

| *Q21* | *Public funding* |
| --- | --- |

Description: Whether the provisioning service has received public funding
Main issues: This information is almost never furnished by the provisioning service
Guidelines for the coding: Select "yes" or "partly" if this information is provided in the website. Select "none" if the information is not given.
Response categories:

1. All
2. Partly
3. None/not reported

| *Q22* | *Turnover* |
| --- | --- |

Description: Turnover is the amount received in sales for a stated period (in Italian "fatturato").
Guidelines for the coding: This information might be given in the provisioning service information page, but its publicity depends on national legislation. Select "not reported" if this information is not retrievable.
Open response or "not reported"

| *Q23* | *Number of registered consumers* |
| --- | --- |

Description: The number of consumers registered to the provisioning service website
Guidelines for the coding: This information might be found in the provisioning service information webpage, but it is rarely provided. Select "not reported" if this information is not retrievable.
Open response or "not reported"

| Q24 | *Number of producers* |
|---|---|

Description: The number of producers registered/present in the provisioning service website

Guidelines for the coding: Count manually the number of producers if this information is not immediately available. Select "not reported" if the website does not allow counting the number of producers.

Open response or "not reported"

| Q25 | *Definition of production standards* |
|---|---|

Description: Information on the standards used by the provisioning service to select their products.

Guidelines for the coding: Select "Explicit" if the provisioning service clearly states and gives evidence that all products are selected through a formal protocol; select "None" if this information is not relevant or it is presented in general terms (e.g. "we select our products from local producers who look like they are sort of ok").

Response categories:

1. Explicit
2. None

| Q26 | *Type of food delivery service (value delivery)* |
|---|---|

Description: Information on the ways food is delivered to the consumer

Guidelines for the coding: Select "click and pick up" when the provisioning service just allows food collection and does not ship products. Select "home delivery" when food is only shipped home. Select "both" when consumer can choose

Response categories:

1. Click and pick up
2. Home delivery
3. Both
4. Not reported

| Q27 | *Value creation (value added)* |
|---|---|

Description: Information on why the provisioning service is attractive to customers

Guidelines for the coding: Select “special products” when the provisioning service sells products with particular characteristics (e.g. gluten free, vegan, ethnic). Select “high quality products” when the products are selected because of their high quality (as stated in the website). Select “low prices” when the provisioning service highlights that food products are sold at cheaper prices. Select “convenience” if products are “ready to cook” or “ready to eat.” Select “organic” if all products sold are organic. Multiple answers are now possible.

Response categories (all that apply):

1. Special products (gluten free, vegan, ethnic, nutraceutical)
2. High-quality products
3. Low prices
4. Convenience
5. Organic
6. Other (please specify)

| *Q28* | *Value capture* |
|---|---|

Description: Information on how the provisioning service covers its cost and earns money

Guidelines for the coding: Select “% of sales” if the provisioning service takes a % out of each transaction between the producer and the consumer and this information is clearly reported. Select “advertising” if the provisioning service has ads and banners on the website. Select “fee” if the provisioning service offers an online window shop to producers that pay a monthly/annual fee and this information is clearly reported. Select “price” if the producers state that they gain by applying a higher price to the products they buy from producers. Select “not reported” if this information is not present.

Response categories (all that apply):

1. % of sales
2. Advertising
3. Fee
4. Price
5. Not reported

| Q29 | *Type of payment (PayPal, credit card, cash on delivery)* |
|---|---|

Description: Information on how customers can purchase the products online

Guidelines for the coding: In the website this information is almost always reported. When absent, try to purchase a product and see which options are given. Select "yes" if the type of payment is present. Select "no" if it's not present.

Response categories:

1. Yes
2. No

| Q30 | *Mission statement* |
|---|---|

Description: Whether the provisioning service has a section which describes, even very shortly, the objectives and the mission of the project

Guidelines for the coding: Select "yes" every time a description (even very short) of the mission/objectives is furnished by the provisioning service

Response categories:

1. Yes
2. No

| Q31 | *Summary of the mission* |
|---|---|

Description: Short description of the mission

Guidelines for the coding: English translation of the mission (or a summary)

Open response

| Q32 | *Mobile app, Facebook page/group, Twitter, Snapchat, Instagram, Pinterest, Google+, WhatsApp, Telegram, YouTube channel* |
|---|---|

Description: Social media where the provisioning service is active

Guidelines for the coding: Usually reported at the bottom of the website. Notice: often necessary to type in Google to see whether the provisioning service is active on social media and these are not indicated in the website. Select “yes” if the provisioning service has a page/profile on the social media; select “no” otherwise.

Response categories:

1. Yes
2. No

---

| *Q33* | *Registration required to access content/service* |
|---|---|

Description: The provisioning service allows you to buy food products only with registration or also as a guest

Guidelines for the coding: Select “yes” if you are obliged to register in order to purchase the food products. Select “no” otherwise.

Response categories:

1. Yes
2. No

---

| *Q34* | *Blog with updated information* |
|---|---|

Description: The provisioning service has a blog with articles

Guidelines for the coding: Select “yes” if the provisioning service has a blog with updated posts (less than 1 year old)

Response categories:

1. Yes
2. No

---

| *Q35* | *Presence of videos on social media pages and/or website* |
|---|---|

Description: The provisioning service hosts video related to the service in the webpage

Guidelines for the coding: Select “yes” if there are videos in the website or in the social media pages

Response categories:

1. Yes
2. No

| *Q36* | *Presence of a review on social media pages and/or the website* |
|---|---|

Description: The provisioning service hosts a review system of the provisioning service itself

Guidelines for the coding: Select "yes" if there is a review system in the website or in the social media pages. Select "no" otherwise.

Response categories:

1. Yes
2. No

| *Q37* | *Organic products* |
|---|---|

Description: The provisioning service sells organic products or not

Guidelines for the coding: Select "sells only certified organic products" if all products sold are organic. Select "does not sell organic products" if there are no organic products at all. Select "both" if there is at least one organic product sold.

If you are unsure how to look for this information use the search box and type "organic": if at least one result is displayed, tick "sells both organic and non-organic products."

Response categories:

1. Sells only certified organic products
2. Does not sell certified organic products
3. Sells both organic and non-organic products
4. Other (please specify)

| *Q38* | *Statements regarding food miles* |
|---|---|

Description: The provisioning service states that products are shipped only within few miles for environmental reasons

Guidelines for the coding: Select "yes" if this is clearly identifiable/stated in the website. Select "no" otherwise.

Response categories:

1. Yes
2. No

| *Q39* | *Statements regarding local food* |
|---|---|

Description: The provisioning service sells local products

Guidelines for the coding: Select "yes" if products come from a specific region/area (BUT NOT NATION) and/or it is clearly stated in the website. Select "no" otherwise.

Response categories:

1. Yes
2. No

| *Q40* | *Statements regarding wellbeing/health* |
|---|---|

Description: The provisioning service sells products which are beneficial for the wellbeing/health of the customer

Guidelines for the coding: Select "yes" only if this is clearly stated. Select "no" otherwise.

Response categories:

1. Yes
2. No

| *Q41* | *Statements regarding environmental protection* |
|---|---|

Description: The provisioning service sells/ships products keeping an eye on environmental concerns

Guidelines for the coding: Select "yes" only when this commitment is clearly stated in the website. Select "no" otherwise.

Response categories:

1. Yes
2. No

| *Q42* | *Statements regarding equity/fair trade* |
|---|---|

Description: The provisioning service values fair trade, working condition, and producers

Guidelines for the coding: Select "yes" if it is clearly stated in the website that they care for equity and fair trade condition. Select "no" otherwise.

Response categories:

1. Yes
2. No

| *Q43* | *Statements regarding seasonal* |
|---|---|

Description: The provisioning service values seasonality of the food products sold

Guidelines for the coding: Select "yes" if this is clearly stated in the webpage. Select "no" otherwise.

Response categories:

1. Yes
2. No

| *Q44* | *Statements regarding cultural heritage/skills/traditions* |
|---|---|

Description: The provisioning service values artisanal products, traditions, and producers respecting particular crafting processes

Guidelines for the coding: Select "yes" if this is clearly stated in the webpage. Select "no" otherwise.

Response categories:

1. Yes
2. No

| *Q45* | *Statements regarding food and nutrient needs* |
|---|---|

Description: The provisioning service sells products for a particular target group (e.g. celiac or vegan or lactose-free)

Guidelines for the coding: Select "yes" if the provisioning service clearly targets at least one group (e.g. gluten-free online shop) or if the provisioning service highlights this in the website (e.g. there is a category of food products named, e.g., "vegan" or "gluten free" or "lactose free"). Select "no" otherwise.

Response categories:

1. Yes
2. No

| *Q46* | *Statements regarding food waste* |
|---|---|

Description: The provisioning service tries to avoid food waste and/or adopts systems to reduce food waste. Guidelines for the coding: select "yes" only if clearly stated in the webpage and you can read the words "food waste" in your language. Select "no" otherwise.

Response categories:

1. Yes
2. No

| *Q47* | *Statements regarding pesticide free* |
|---|---|

Description: The provisioning service sells products that are pesticide free

Guidelines for the coding: Select "yes" if this commitment is clearly stated in the website. Select "no" otherwise.

Response categories:

1. Yes
2. No

| *Q48* | *Statements regarding GMOs free* |
|---|---|

Description: The provisioning service sells or values products that are GMOs free

Guidelines for the coding: Select "yes" if this is clearly stated in the website. Select "no" otherwise.

Response categories:

1. Yes
2. No

# Appendix B: Interviewing Schedule

Version (A): If the photo food dairy is detailed and contains enough material, then the interview can begin by asking about the photos and work in the questions of the interview guide.

Version (B): If the photo food diary is useful but not detailed enough to structure the interview around, then begin with the questions and integrate questions about the photos when relevant.

1. **Introduction of Interviewee**

Would you please introduce yourself briefly and tell us something about you and your household?

(If version A)

Let's have a look at your pictures—please describe your daily activities regarding food acquisitioning and eating!

2. **Food Shopping/Provisioning**

Can you tell me about the last time you shopped groceries?

- When/what/where/who?
- (if captured in photo diary use the material)



A. Dulsrud, F. Forno (eds.), *Digital Food Provisioning in Times of Multiple Crises*, Consumption and Public Life,
https://doi.org/10.1007/978-3-031-46323-5

- How did you plan that shopping trip?
- shopping performance—at the store (online/offline)—variations
- after shopping—bringing grocery home/receiving grocery
- after shopping evaluations of shopping

Can you give me another example of a recent shopping trip?

- (see above)

What would you say is a typical (grocery) shopping week for you?

How would you say that [x type of shopping] is different from [y type of shopping]?

Do you have any other way of acquiring food (or drinks)? Could you tell me?

- (give examples, sharing, growing your own, etc.)

3. **Eating**

Can you tell me about the last meal that you had?

- When/what/where/who?
- (if captured in photo diary use the material)

Can you give me another example of a recent meal?

- When/what/where/who?
- (if captured in photo diary use the material)

Can you tell me about breakfast/lunch/dinner/snacks?

- When/what/where/whom?
- (if the informants do not eat breakfast/lunch/dinner/snacks ask why)

Would you characterize this (breakfast/lunch/dinner/snack) as a "proper" meal? What would a proper meal entail? (possible directions: ingredients, drinks, social setting, time frame, etc.)

4. **Cooking**

Can you tell me about the last meal that you cooked/prepared?

- When/what/where/whom?
- (if captured in photo diary use the material)

Can you give me another example of a recent meal you cooked/prepared?

- When/what/where/whom?
- (if captured in photo diary use the material)

Who does the cooking in the household? Does this change according to schedule? Has it changed over time?

Would you characterize yourself as a skillful cook? How/when/where/from whom did you learn to cook?

5. **Storing Food**

Where and how do you store your food (and drinks)? Can you show me please?

- Where is it placed—cupboards, fridge, other?
- When is it placed there?
- Principle of organization

6. **Disposing of Food**

What was the last food item that you threw away?

(Ask to see the garbage bin and talk about why it was wasted, by whom, what it was intended for) Photograph garbage bin

Do you have any leftovers in your fridge now? Can you show me, please? (Ask to see the fridge and talk about the leftovers in the fridge. What does it consist of, why was there leftovers, what is the plan for the leftovers, will they be eaten or wasted?) Photograph the leftovers.

What happens with leftovers—how is it preserved or included into new meals?

7. **Platform Specific Questions**

[These questions have to be formulated specific for the platform/pipeline under study. Also when digital these questions could be combined with the digital walkthrough—questions regarding the usage of platforms can and should be asked in every section of the interview]

Can you tell me about the last time you used [Name of the platform]?

- When/what/where/whom?
- (if captured in photo diary use the material)

What did your food provisioning look like before you were using [Name of the platform]?

Which other channels of food provisioning are you using? Can you tell me about those?

Why are you using exactly this platform—what is it offering you?

Further topics to consider:

- frequency, connection to other activities
- influence of money, time, availability of products

8. **End**

Do you think the way you acquisition and prepare food has an influence on matters of sustainability?

If you could change one thing regarding your food acquisitioning and preparing what would that be?

Do you have any other issues or thoughts before ending the interview?

# Appendix C: Online Survey

**Introductory Text:**

*Invitation to Take Part in the Survey*

We invite you to take part in this survey on food practices and online platforms for food acquisition.

This survey is part of the SUSFOOD2 ERA- NET Plateforms project that explores new ways of buying food in five European countries. By answering the following questions, you will help us understand the impact of various types of new online food platforms in everyday food practices.

*European Collaboration Project*

The project is a collaboration between researchers in Norway, Sweden, Ireland, Germany, and Italy. The project is led by Consumption Research of Norway at Oslo Metropolitan University, and the University of Trento has the responsibility for this survey.

*Anonymous Survey*

The data will be collected via Limesurvey by the University of Trento and will be used for research and results published in academic journals focusing on food consumption.

The survey is anonymous. The survey will ask the following characterizing information: gender, age range, size of the area of residence, educational level, economic status, and family composition.



A. Dulsrud, F. Forno (eds.), *Digital Food Provisioning in Times of Multiple Crises*, Consumption and Public Life,
https://doi.org/10.1007/978-3-031-46323-5

The record of your survey responses will not contain any identifying information about you; questions are designed to ensure sufficient anonymity of data. The data will be only treated and presented in anonymized and aggregated form, and only used for scientific purposes.

*Use and Storage of the Data*

The anonymous data you provide will be used by researchers in the Plateforms project. The anonymous data will be stored by the University of Trento until the end of 2022, and thereafter transferred to permanent and anonymous storage according to the university regulations. The anonymous permanently stored data will be available for other researchers for academic work.

*Voluntary to Answer*

It is not compulsory to answer all the questions and you can interrupt the survey at any time. Partial information will still be used.

*Contact Information*

For any information about this survey, you may contact the team responsible for the survey at the following e-mail address: each country inserts its own.

*Consent*

I hereby consent to take part in this survey, and that the anonymous data can be used as described above.

## Part 1—Demographics

*Introductory text*

| P1(int)a | Please introduce yourself |
|---|---|

Code/question/answer options/information notes

| | |
|---|---|
| *Q1* | *Country of residence* |
| q1No | Norway |
| q1Sw | Sweden |
| q1Ir | Ireland |
| q1It | Italy |
| q1Ge | Germany |

| | |
|---|---|
| *Q2* | *Age* |
| q2a | < 30 |
| q2b | 30–44 |
| q2c | 45–59 |
| q2d | > 60 |

| | |
|---|---|
| *Q3* | *Gender* |
| q3f | Female |
| q3m | Male |
| q3o | Other |
| q3n | Prefer not to say |

| | |
|---|---|
| *Q4* | *How many people live in your household? Indicate the number for each category represented in your household in the box on the right* |
| q4a | Adults [state number] |
| q4b | Children (under 6 years of age) [state number] |
| q4c | Children (6–17 years of age) [state number] |

| | |
|---|---|
| *Q5* | *Do you live in a rural or urban area?* |
| q5r | Rural area |
| q5t | Town with less than 25,000 people |
| q5us | Small city with 25,000 to 200,000 people |
| q5ub | City with more than 200,000 people |

| | |
|---|---|
| *Q6* | *Which of these best describes your level of education?* |
| q6a | No formal education |
| q6b | Primary education |
| q6c | Secondary education |
| q6d | More than secondary school |
| Q6e | University-level degree (B.A., M.A., Ph.D.) |

| | |
|---|---|
| *Q7* | *What is your main occupation (choose the one that best describes you)?* |
| q7f | Full-time paid work, more than 30 hours per week |
| q7p | Part-time paid work, between 15 and 29 hours per week |
| q7u | Unemployed |
| q7o | Other (student, homemaker, retired, etc.) |

| | |
|---|---|
| *Q8* | *On an income scale on which 1 indicates the lowest income group and 10 the highest income group in your country. We would like to know in what group your household is. Please specify the appropriate number, counting all wages, salaries, pensions, and other incomes that come in.* |
| Q8a | 1–10 Highest income |

| | |
|---|---|
| SQ8A | People sometimes describe themselves as belonging to the working class, the middle class, or the upper or lower class. Would you describe yourself as belonging to the<br>• Upper class<br>• Upper middle class<br>• Lower middle class<br>• Working class<br>• Lower class |
| SQ8B | How has the financial situation of your household changed over the last 12 months?<br>• Got a lot better<br>• Got a little better<br>• Stayed the same<br>• Got a little worse<br>• Got a lot worse<br>• Don't know |
| *SQ8C* | *How do you expect the financial position of your household to change over the next 12 months?*<br>• Get a lot better<br>• Get a little better<br>• Stay the same<br>• Get a little worse<br>• Get a lot worse<br>• Don't know |

## Part 2—Experience with Online Platforms for Food Acquisition

Introductory Text

Food groceries: In this survey, food groceries are understood as all food products usually available in a grocery store. This includes, among others dairy products, eggs, meat, seafood, oils, bread, fruits, vegetables, beer, soft drinks, coffee, tea, snacks, and chocolates.

Online platforms for food acquisition: In this survey, online platforms for food acquisition are understood as all websites, mobile applications, or similar, where you can order or purchase food groceries, including online ordering or purchases from grocery stores, specialty stores, and various services and groups with home delivery or food collection points. It also includes various types of online food subscription services. We do not include online services for booking at the restaurant, getting takeout, or home delivery of restaurant meals in online platforms for food acquisition.

Online food buyer communities: Organized by consumers, informal groups, or non-commercial associations

## Subgroup A

| Code/question | items/answer options/information notes |
|---|---|

*Q9* *Which of the following online platforms have you used for food acquisition in the last year?*
Country specific options_______
Select the name of the platform you have used_______

*SQ9* *Approximately, when did you start using online platforms for food acquisition on a regular basis?*
- Before 2010
- 2010–2014
- 2015–2019
- 2020
- Do not use on a regular basis

| | |
|---|---|
| *Q10* | *Approximately, how often does your household order, purchase, or receive food groceries from the following online platforms for food acquisition?* |
| q10a | Online supermarkets |
| q10b | Online food stores specializing in one or a few products categories or consumer segments |
| q10c | Online food subscription services with recipes and the needed meal ingredients |
| q10d | Other types of online food subscription services |
| q10e | Online platform to buy directly from producers (e.g. food assembly) |
| q10f | Online consumer groups or associations |
| q10g | Other |
| sq10a | Weekly or more often |
| sq10b | Once every two weeks |
| sq10c | Once every three to four weeks |
| sq10d | Once every two to three months |
| sq10e | Less than once every three months |
| sq10f | Never |

| | |
|---|---|
| *Q11* | *How do you receive most of the food groceries you order or purchase on online platforms for food acquisition? Rank the three alternatives, where 1 is most often and 3 is least often.* |
| q11a | I receive them at my address |
| q11b | I receive them from a food collection point |
| q11c | Other |

| | |
|---|---|
| *Q12* | *How often do you buy the following food items on online platforms for food acquisition?* |
| q12a | Dairy products and eggs |

| | |
|---|---|
| q12b | Meat and meat products |
| q12c | Fish and seafood |
| q12d | Bread and bakery products |
| q12e | Fresh fruits and vegetables |
| q12f | Oils |
| q12g | Canned/bottled food products |
| q12h | Seeds, dried fruits, and pulses |
| q12i | Non-alcoholic beverages |
| q12j | Alcoholic beverages |
| q12k | Coffee and tea |
| q12l | Snacks and confectionery |
| q12m | Pasta, rice, and grains |
| Q12n | Products for special dietary needs |
| sq12a | Weekly or more often |
| sq12b | Once every two weeks |
| sq12c | Once every three to four weeks |
| sq12d | Once every two to three months |
| sq12e | Less than once every three months |
| sq12f | Never |

| | |
|---|---|
| *Q13* | *Approximately, how much did you spend on online platforms in the last 3 months?* |
| q13a | Less than 50 euros |
| q13b | 50–99 euros |
| q13c | 100–499 euros |
| q13d | 500–1000 euros |
| q13e | More than 1000 euros |

## Subgroup B

Introductory Text

| | |
|---|---|
| P2B(int)a | Many consumers have experienced a change in their food purchasing practices due to the COVID-19 pandemic. We will now ask you to recall your grocery purchasing in the months preceding the COVID-19 pandemic and compare them with your grocery purchases during the COVID-19 pandemic. We would also like you to indicate your online purchase expectations over the coming years. |
| Code/ question | items/answer options/information notes |

| | | |
|---|---|---|
| *Q14* | *What share of your food grocery shopping was done online in the months before the COVID-19 pandemic?* | |
| q14a | NOTHING 0 … 10 EVERYTHING | 10-point scale |

| | | |
|---|---|---|
| *Q15* | *What share of your food grocery shopping was done online in the months since the start of the COVID-19 pandemic?* | |
| q15a | NOTHING 0 … 10 EVERYTHING | 10-point scale |

| | |
|---|---|
| *Q16* | *Look a few years ahead, when the COVID-19 pandemic is over. Compared to now, do you think the share of your food grocery shopping done online will increase, remain unchanged, or decrease?* |
| q16a | Increase |
| q16b | Remain unchanged |
| q16c | Decrease |

## Part 3—What Would Your Ideal Online Platform for Food Acquisition Be?

### Introductory Text

| | |
|---|---|
| P3(int)a | Imagine that in a few years' time you could choose between many online platforms for food acquisition in your area. Which factors would be most and least important for your choice of online platforms for food acquisition? |

You will now get six ranking questions. Each question presents three factors, and for each question you are asked to say which of the three factors would be most and least important for your future choice of online platforms for food acquisition.

| Code/question | items/answer options/information notes |
|---|---|

| | |
|---|---|
| *Q17* | *Of the following three factors, which factors would be the most and least important in your future choice of online platforms for food acquisition?* |
| X1 | They have a wide range of organic food |
| X2 | They have a wide range of local products |
| X3 | They have a wide range of products from small-scale producers |
| X4 | They have a website that is easy to use |
| X5 | They provide recipe suggestions for the foods they sell |
| X6 | They have the same product range as a supermarket |
| X7 | They provide products without disposable packaging |
| X8 | They provide fruits and vegetables that are in season |
| X9 | They have strict quality standards for the products they sell |
| X10 | They deliver to your home |
| X11 | They deliver at a food collection point |
| X12 | They deliver within two days |
| X13 | They provide opportunities to socialize with other users |

| | |
|---|---|
| X14 | They provide opportunities to meet producers face to face |
| X15 | They provide opportunities to learn about products from other customers |
| X16 | They have a subscription service with the same payment each time |
| X17 | They charge based on how much you buy each time |
| X18 | They have no delivery charges |

## Part 4—What Kind of Food Consumer Are You?

### Introductory Text

| | |
|---|---|
| P4(int)a | How well do the following statements describe you as a food consumer? 7-point scale from 1 = Do not describe me at all to 7 = Describe me very well. |
| Code/ question | items/answer options/information notes |

| | |
|---|---|
| *Q18* | *Please indicate how each statement reflects you as a consumer* |
| Y1+ | I usually use a shopping list |
| Y2+ | I usually walk, cycle, or take public transport for my food shopping |
| Y3+ | I mainly buy fresh fruits and vegetables that are in season |
| Y4+ | I prefer domestically produced food |
| Y5+ | I prefer locally produced food |
| Y6+ | I eat at least five portions of fruits and vegetables a day |
| Y7+ | I try to avoid consuming red meat |
| Y8+ | I eat vegetarian meals at least every second day |
| Y9+ | I prefer organic food |
| Y10+ | I have a good overview of the food I have at home |
| Y11+ | I sometimes eat food even if it has past the best before date |
| Y12+ | I store and use leftover food |
| Y13+ | I try to avoid creating food waste |
| Y14+ | I cook most of my meals from scratch |
| Y15+ | I value quality over price |
| Y16+ | I prefer products where I can be sure that the producer has received a fair price |

| | |
|---|---|
| *Q19* | *Please indicate how each statement reflects you as a consumer* |
| Y1- | I frequently buy food products I had not planned to buy |
| Y2- | I frequently drive to the place I do the food shopping |
| Y3- | I am not very interested in the seasonality of products |
| Y4- | I rarely check the country of origin of the food I buy |
| Y5- | I am not very interested in locally produced food |
| Y6- | I eat very little fruits and vegetables |
| Y7- | I have meat-based meals most days |
| Y8- | I seldom eat vegetarian meals |
| Y9- | I rarely buy organic food |

| | |
|---|---|
| Y10- | I often find food in my kitchen that I had forgotten about |
| Y11- | I always throw away food that is past the best before date |
| Y12- | I rarely eat leftover food |
| Y13- | I sometimes find food in my kitchen that has gone off |
| Y14- | I frequently buy ready-made food |
| Y15- | I usually choose the cheapest option |
| Y16+ | I seldom choose Fairtrade products |

# GPSR Compliance

*The European Union's (EU) General Product Safety Regulation (GPSR) is a set of rules that requires consumer products to be safe and our obligations to ensure this.*

*If you have any concerns about our products, you can contact us on ProductSafety@springernature.com*

In case Publisher is established outside the EU, the EU authorized representative is:

Springer Nature Customer Service Center GmbH
Europaplatz 3
69115 Heidelberg, Germany

Kolibri 360 GmbH
Ferdinand-Jühlke-Straße 7
99095 Erfurt, Deutschland
produktsicherheit@kolibri360.de